Contents

Introduction

Cambridge Advanced Level Mathematics has been written especially for the OCR Mathematics specification. It consists of one book or half-book corresponding to each module. This book is the first Discrete Mathematics module, D1.

The books are divided into chapters roughly corresponding to specification headings. Occasionally a section includes an important result that is difficult to prove or outside the specification. These sections are marked with an asterisk (*) in the section heading, and there is usually a sentence early on explaining precisely what it is that the student needs to know.

It is important to recognise that, while every effort has been made by the author and by OCR to make the books match the specification, the books do not and must not define the examination. It is conceivable that questions might be asked in the examination, examples of which do not appear specifically in the books.

Occasionally within the text paragraphs appear in *this type style*. These paragraphs are usually outside the main stream of the mathematical argument, but may help to give insight, or suggest extra work or different approaches.

Numerical work is presented in a form intended to discourage premature approximation. In ongoing calculations inexact numbers appear in decimal form like 3.456..., signifying that the number is held in a calculator to more places than are given. Numbers are not rounded at this stage; the full display could be either 3.456 123 or 3.456 789. Final answers are then stated with some indication that they are approximate, for example '1.23 correct to 3 significant figures'.

There are plenty of exercises, and each chapter contains a Miscellaneous exercise which includes examination questions. Most of these questions were set in OCR examinations; some were set in AQA examinations. Questions which go beyond examination requirements are marked by an asterisk. At the end of the book there is a set of Revision exercises and two practice examination papers. The author thanks Jan Dangerfield, the OCR examiner who contributed to these exercises, and who also read the book very carefully and made many extremely useful and constructive comments.

The author thanks OCR and Cambridge University Press for their help in producing this book. However, the responsibility for the text, and for any errors, remains with the author.

Discrete Mathematics 1

Stan Dolan

Series editor Hugh Neill

CAMBRIDGE UNIVERSITY PRESS

PUBLISHED BY THE PRESS SYNDICATE OF THE UNIVERSITY OF CAMBRIDGE
The Pitt Building, Trumpington Street, Cambridge, United Kingdom

CAMBRIDGE UNIVERSITY PRESS
The Edinburgh Building, Cambridge CB2 2RU, UK
40 West 20th Street, New York, NY 10011-4211, USA
477 Williamstown Road, Port Melbourne, VIC 3207, Australia
Ruiz de Alarcón 13, 28014 Madrid, Spain
Dock House, The Waterfront, Cape Town 8001, South Africa

http://www.cambridge.org

First published 2000
Fifth printing 2002

Printed in the United Kingdom at the University Press, Cambridge

Typefaces Times, Helvetica   *Systems* Microsoft® Word, MathType™

A catalogue record for this book is available from the British Library

ISBN 0 521 78610 X paperback

1 Algorithms

This chapter looks at the meaning of 'Discrete Mathematics' and introduces some algorithms. When you have completed it you should

- know what an algorithm is
- be able to apply the algorithms known as Bubble Sort, Shuttle Sort and First-Fit
- know how to define the size of a problem, and the efficiency and order of an algorithm.

1.1 What is Discrete Mathematics?

You will have already met, in Statistics, the distinction between continuous and discrete data. Continuous data can take any value in a numerical range: measurements of height, weight and time all produce continuous data. Discrete data can only take values which are strictly separated from each other: measurements of the number of children in a family or the number of letters in a word are discrete data which take only whole-number values.

In the 17th century, Sir Isaac Newton and other leading mathematicians started the development of calculus, which deals specifically with continuous data, and graphs which are generally smooth. Discrete Mathematics deals only with branches of mathematics which do *not* employ the continuous methods of calculus.

However, the distinction between continuous and discrete sometimes becomes blurred. For example, computers essentially deal in Discrete Mathematics, because they hold numbers using sequences of 1s and 0s, and can only hold a finite amount of information. However, advanced computers can work to a very high degree of accuracy, and can do very good approximations to continuous mathematics. They can give approximate solutions to equations which otherwise could not be solved.

Computer screens are divided into 'pixels' (the word is a contraction of 'picture elements'), and so computer and TV screens are essentially discrete devices. However, because the discrete pixels are so small, the images on the screen appear continuous.

But all this is only part of the definition of Discrete Mathematics. It is also widely (but not universally) accepted that Discrete Mathematics is restricted to branches of mathematics whose development has mainly been in the 20th century. It is no coincidence that its importance and application have arisen in the same period of history as the development of computers.

Part of working with computers is the idea of a procedure, or 'algorithm', to solve a problem. You probably know an algorithm which enables you to find the answer to a long multiplication given the two numbers you wish to multiply. Algorithms form a substantial part of Discrete Mathematics. In this course, most of the algorithms will be topics related to the best use of time and resources. These have applications in industry, business, computing and in military matters.

1.2   Following instructions

An algorithm is a sequence of instructions which, if followed correctly, allows anyone
to solve a problem.

The mathematics problems studied in school tend to be those for which previous generations
of mathematicians have already worked out the appropriate sequences of instructions. For
example, consider this algorithm for finding the median of a set of numbers.

Find the median	*Example* 12, 2, 3, 8, 2, 4
Step 1 Arrange the numbers in ascending order.	2 2 3 4 8 12
Step 2 Delete the end numbers.	2 3 4 8
Step 3 Repeat Step 2 until only one or two numbers remain.	3 4
Step 4 The median is the number that remains, or the average of the two numbers that remain.	3.5

It is especially important to think of mathematical procedures as sequences of precise
instructions when you are programming a computer to solve a problem. Computer programs
are algorithms written in a language which a computer can interpret.

Other types of everyday algorithm include cookery recipes, explanations on how to set up
video recorders, and assembly instructions for flat-pack furniture. The following
paragraph, from some instructions recently followed by the author, illustrates some of the
advantages and disadvantages of algorithmic methods.

> From Bag 46 take one $\frac{3}{8}$" × $2\frac{1}{2}$" Hex Head Bolt and one $\frac{3}{8}$" Nyloc Nut. Insert the
> bolt through the bottom hole of the bracket on part 1050 and through the hole in
> part 1241. Attach the Nyloc Nut finger tight.

Providing the instructions are sufficiently precise, you can carefully work through an
algorithm such as this one without needing to fully understand how everything fits
together. Similarly, you can follow mathematical algorithms by rote, without
understanding the process.

However, if you do understand a process then you can adapt the basic algorithm to
special features of the problem, and thereby solve the problem more efficiently. Just as
the author eventually stopped needing detailed instructions on how to attach parts
together with appropriate-sized nuts and bolts, so you would not need to follow the
algorithm slavishly if asked to find the median of $\overbrace{2, 2, 2, \ldots, 2}^{1000 \text{ numbers}}$.

> An **algorithm** is a finite sequence of instructions for solving
> a problem. It enables a person or a computer to solve the
> problem without needing to understand the whole process.

You might wonder why the word 'finite' is necessary. The reason is that there are processes which are essentially infinite, like finding the sum of a series such as

$$1 + \tfrac{1}{2} + \tfrac{1}{4} + \tfrac{1}{8} + \dots$$

by adding successive terms to the 'sum so far'. This never ends, and is not an algorithm.

In this book you will learn some algorithms which have been developed to solve particular problems in Discrete Mathematics. You will need to know how to carry these algorithms out by hand, although most real-world applications involve so many steps that they require the use of a computer. You will also need to have some idea of why the methods work.

1.3 Sorting algorithms

Any collection of data, such as a telephone directory, is only of value if information can be found quickly when needed. Alongside the development of computer databases, many algorithms have been developed to speed up the modification, deletion, addition and retrieval of data. This section will consider just one aspect of this, the sorting of a list of numbers into numerical order.

There is no single 'best' algorithm for sorting. The size of the data set, and how muddled up it is initially, both affect which algorithm will sort the data most efficiently.

Bubble Sort

This algorithm is so called because the smaller numbers gradually rise up the list like bubbles in a glass of lemonade. The algorithm depends upon successive comparisons of pairs of numbers, as follows.

- Compare the 1st and 2nd numbers in the list, and swap them if the 2nd number is smaller.
- Compare the 2nd and 3rd numbers and swap if the 3rd is smaller.
- Continue in this way through the entire list.

Consider the application of this procedure, called a **pass**, to the list of numbers

   5, 1, 2, 6, 9, 4, 3.

The numbers are first placed vertically in the left column. After each comparison the list is rewritten to the right. Fig. 1.1 shows one pass of Bubble Sort.

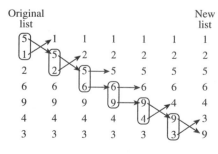

Fig. 1.1

You can see that this pass has required 6 comparisons and 4 swaps.

The result of this pass through the list is that the numbers 1, 2, 4 and 3 (the bubbles) have each moved up one place. The other numbers have either stayed in place or moved down. In particular, you should be able to see that the largest number (in this case the 9) will always move to the bottom.

The complete Bubble Sort algorithm can be written as follows.

Bubble Sort

Step 1	If there is only one number in the list then stop.					

Original list	1st pass	2nd pass	3rd pass	4th pass	5th pass
5	1	1	1	1	1
1	2	2	2	2	2
2	5	5	4	3	3
6	6	4	3	4	4
9	4	3	5	5	5
4	3	6	6	6	6
3	9	9	9	9	9

Step 2 Make one pass down the list, comparing numbers in pairs and swapping as necessary.

Step 3 If no swaps have occurred then stop. Otherwise, ignore the last element of the list and return to Step 1.

Fig. 1.2

Each pass alters the list of numbers as in Fig. 1.2, and the list ends up in order. The numbers under the 'steps' are the ones that are ignored.

Table 1.3 shows the numbers of swaps and comparisons which are required at each pass.

	1st pass	2nd pass	3rd pass	4th pass	5th pass	Totals
Comparisons	6	5	4	3	2	20
Swaps	4	2	2	1	0	9

Table 1.3

One disadvantage of Bubble Sort is that once the data have been sorted, another complete pass through the data is necessary to ensure that the sorting has been finished. The next algorithm partially overcomes this problem.

Shuttle Sort

This algorithm is so called because numbers can move up more than one place in a pass.

Shuttle Sort

1st pass Compare the 1st and 2nd numbers in the list and swap if necessary.

2nd pass Compare the 2nd and 3rd numbers in the list and swap if necessary. If a swap has occurred, compare the 1st and 2nd numbers and swap if necessary.

3rd pass Compare the 3rd and 4th numbers in the list and swap if necessary. If a swap has occurred, compare the 2nd and 3rd numbers, and so on up the list.

And so on, through the entire list.

The results of successive passes of Shuttle Sort on the list

5, 1, 2, 6, 9, 4, 3

are shown in Fig. 1.4. The numbers above the stepped line are those which have been compared at each pass.

Original list	1st pass	2nd pass	3rd pass	4th pass	5th pass	6th pass
5	1	1	1	1	1	1
1	5	2	2	2	2	2
2	2	5	5	5	4	3
6	6	6	6	6	5	4
9	9	9	9	9	6	5
4	4	4	4	4	9	6
3	3	3	3	3	3	9

Fig. 1.4

Table 1.5 shows the numbers of swaps and comparisons which are required at each pass of Shuttle Sort.

	1st pass	2nd pass	3rd pass	4th pass	5th pass	6th pass	Totals
Comparisons	1	2	1	1	4	5	14
Swaps	1	1	0	0	3	4	9

Table 1.5

Shuttle Sort has involved the same number of swaps as Bubble Sort, but far fewer comparisons: 14 as opposed to 20.

Exercise 1A

1 (a) Apply Bubble Sort to the reverse-ordered list 5, 4, 3, 2, 1. Keep a count of the number of comparisons and swaps.

(b) Apply Shuttle Sort to the list 5, 4, 3, 2, 1. Again, keep a count of the number of comparisons and swaps.

(c) Compare the two sorting algorithms for lists in reverse order.

2 (a) Apply Bubble Sort to a list of 6 numbers. What is the maximum possible number of comparisons and swaps that would need to be made for a list of 6 numbers?

(b) Generalise your answer to part (a) for a list of n numbers.

3 Apply Shuttle Sort to the list 4, 1, 6, 8, 2. Show the result of each pass and keep a count of the number of comparisons and swaps.

4 Another sorting algorithm, called the Interchange algorithm, is defined as follows.
Step 1 If there is only one number in the list then stop.
Step 2 Find the smallest number in the list and interchange it with the first number.
Step 3 Ignore the first element of the list and return to Step 1.

(a) Write down your own sub-algorithm to 'find the smallest number in a list'. How many comparisons are needed when applying your algorithm to a list containing n numbers?

(b) How many comparisons and swaps are needed when applying the Interchange algorithm to the list 5, 4, 3, 2, 1?

5 Here are two algorithms. In each case, find the output if $m = 4$ and $n = 3$, and decide whether the algorithm would still work if either or both m and n were negative.

   (a) **Step 1**   Read the positive integers m and n.
            Step 2   Replace m by $m-1$, and n by $n+1$.
            Step 3   If $m > 0$, go to Step 2. Otherwise write n.

   (b) **Step 1**   Read the positive integers m and n.
            Step 2   Let $p = 0$.
            Step 3   Replace m by $m-1$, and p by $p+n$.
            Step 4   If $m > 0$, go to Step 3. Otherwise write p.

1.4   The order of an algorithm

The following definitions are useful in deciding how well an algorithm performs its task.

The **efficiency** of an algorithm is a measure of the 'run-time' for the algorithm. This will often be proportional to the number of operations which have to be carried out.

The **size** of a problem is a measure of its complexity. In the case of a sorting algorithm, this is likely to be the number of numbers in the list.

The **order** of an algorithm is a measure of the efficiency of the algorithm as a function of the size of the problem.

Example 1.4.1
Determine the order of the Bubble Sort algorithm.

Apply the algorithm to a list with n elements. The maximum possible number of comparisons (and of swaps) is

$$(n-1)+(n-2)+\ldots+1 = \tfrac{1}{2}(n-1)n.$$

In this case, $\tfrac{1}{2}(n-1)n$ is a measure of efficiency and n is a measure of size. Efficiency is therefore a quadratic function of size and the Bubble Sort algorithm is said to be 'of quadratic order', or 'of order n^2'.

Table 1.6 gives some examples of different orders of algorithm.

Algorithm	Size	Efficiency	Order
A	n	$5n$	n, or linear
B	n	$n^2 + 7n$	n^2, or quadratic
C	n	$2n^3 - 3n$	n^3, or cubic

Table 1.6

Algorithms A, B and C all have polynomial orders. In general, algorithms which have polynomial orders are considered satisfactory for tackling large problems, because a computer can usually complete the solution in a reasonable amount of time.

Example 1.4.2

A computer is programmed with an algorithm of quadratic order. Given that the computer takes two seconds to solve a problem of size 15, estimate the time it would take to solve a problem of size 150.

The run-time is roughly proportional to n^2, so you can estimate the time as follows.

Since n is multiplied by 10, n^2 is multiplied by 100. So the time is multiplied by 100.

Therefore the new time is 200 seconds, which is just under 3.5 minutes.

Some algorithms have orders which are not polynomials. Some that occur quite frequently have efficiencies which depend upon

$$n \times (n-1) \times (n-2) \times \ldots \times 1,$$

where n is the size. This expression is called 'factorial n' (see P2 Section 3.4), and it is written as $n!$. Other algorithms may depend upon 2^n.

The functions $n!$ and 2^n are similar types of function, called 'exponential'. Algorithms of **exponential** order are not regarded as ideal for large problems, because they take too long. Compare the following problem with Example 1.4.2.

Example 1.4.3

A computer is programmed with an algorithm of exponential order 2^n. Given that the computer takes two seconds to solve a problem of size 15, estimate the time it would take to solve a problem of size 150.

The run-time t seconds is roughly proportional to 2^n, so $t = k \times 2^n$.

When $n = 15$, $t = 2$, so $k = \dfrac{2}{2^{15}}$.

Therefore, when $n = 150$, $t = k \times 2^{150} = \dfrac{2}{2^{15}} \times 2^{150} = 2 \times 2^{135} = 8.71\ldots \times 10^{40}$.

The new time is about 8.7×10^{40} seconds, which is roughly 2.8×10^{33} years!

Modern cryptography, which is very important in business and finance, as well as for military purposes and national security, is based on codes which can be decoded using standard algorithms. However, the only known algorithms are of exponential order, and so the codes are regarded as being sufficiently secure.

1.5   Packing algorithms

Ferry companies have to make the best possible use of the space available on their ships, and they make considerable efforts to ensure that they pack vehicles as efficiently as possible. Similar packing problems occur in warehouses. Companies do not want to waste money having larger warehouses simply because racks have been stacked inefficiently, so modern warehouses have computerised systems to organise storage.

One of the many systematic methods for packing is called the First-Fit algorithm. Although this rarely leads to the 'best' solution it does have the advantage of being simple.

First-Fit algorithm
Place each object in turn in the first available space in which it will fit.

Example 1.5.1
A small ferry has three lanes, each 25 metres long. The lengths in metres of the vehicles in the queue, in the order in which they are waiting, are

3   5   4   3   14   5   9   3   4   4   4   3   11.

(a) Use the First-Fit algorithm to load the ferry with vehicles.
(b) Find a more efficient packing of the ferry in this particular case.

(a) According to the algorithm, the vehicles are packed as follows.

Lengths	3	5	4	3	14	5	9	3	4	4	4	3	11
Lane 1	3	5	4	3		5		3					
Lane 2					14		9						
Lane 3									4	4	4	3	

Notice that the 11-metre vehicle at the end does not fit.

(b) One possibility is to put the 9-metre vehicle into lane 3. The 11-metre vehicle then fits in lane 2.

Exercise 1B

1 A carpenter is cutting pieces of timber from 8-foot lengths. The lengths (in feet) required are 1, 4, 2, 4, 3, 2, 5, 2.

(a) How many 8-foot lengths need to be cut if the First-Fit algorithm is used?

(b) How could the timber be cut more efficiently?

(c) In answering this question, what simplifying assumption have you made about cutting the timber?

2 A computer takes 10^{-6} seconds to perform a certain algorithm on a list with only one number. Estimate how long the algorithm will take to perform the algorithm on a list with 1000 numbers if the order of the algorithm is

(a) n,   (b) n^2,   (c) n^3,   (d) 2^n.

3 One attempt to improve the First-Fit algorithm is called the First-Fit Decreasing algorithm. For this, the objects to be packed are first ordered in decreasing size, and then First-Fit is applied. Apply the First-Fit Decreasing algorithm to the ferry problem of Example 1.5.1.

4 Sometimes the First-Fit algorithm is, by chance, better than the First-Fit Decreasing algorithm. Suppose two storage bins of height 200 cm are used for boxes of a standard length and width but varying heights. Find heights of six boxes which can be arranged in such a way that First-Fit, but not First-Fit Decreasing, will enable you to completely fill the bins.

5 The following algorithm divides an $m \times n$ rectangle into squares.

Step 1 If $m = n$ then stop.

Step 2 If $n > m$ then swap m and n.

Step 3 Divide the rectangle into an $n \times n$ square and an $(m-n) \times n$ rectangle.

Step 4 Replace m by $m - n$.

Step 5 Go to Step 1.

For example, the figure shows how a 3×5 rectangle is split up into four squares.

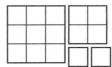

(a) Complete this table of values of m and n for the algorithm applied to the 3×5 rectangle.

Step	1	2	4	1	2	4	1	2	...
m	3	5							
n	5	3							

(b) Into how many squares does the algorithm divide a 5×6 rectangle? Find a division of a 5×6 rectangle into fewer squares.

1.6 Flow diagrams

A flow diagram is a pictorial representation of an algorithm. Differently shaped boxes are used for different types of instruction. Fig. 1.7 shows you which instructions go into which boxes.

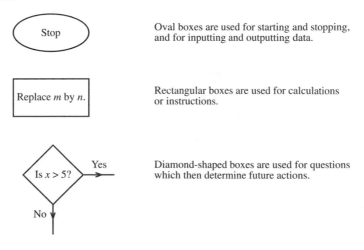

Oval boxes are used for starting and stopping, and for inputting and outputting data.

Rectangular boxes are used for calculations or instructions.

Diamond-shaped boxes are used for questions which then determine future actions.

Fig. 1.7

The algorithm given in
Section 1.2 for finding the
median of a set of numbers
can be represented by the
flow diagram in Fig. 1.8.

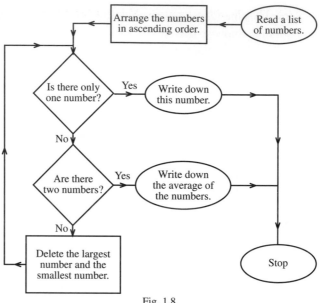

Fig. 1.8

1.7   Notation for algorithms

In Fig. 1.7, the rectangle contained the instruction 'Replace m by n'. Think of m and n as labels of pigeon-holes, each of them containing a number. This instruction means take the number which is in pigeon-hole n and put it into pigeon-hole m, replacing the number which is already there. The notation used in this book for this instruction will be $m := n$.

Similarly, $m := 2$ means put the number 2 into pigeon-hole m; and $m := m - 1$ means take the number already in pigeon-hole m, subtract 1 from it, and put the result back into pigeon-hole m.

These pigeon-holes are usually called **stores**.

Example 1.7.1

An algorithm has a flow diagram which is shown on the opposite page.
(a) What is the output if $N = 57$?
(b) What has this algorithm been designed to do?

(a) After successive passes around the flow diagram, the values of N, R and the numbers written down so far are as shown in the table.

Pass	N	R	Written down
1	28	1	1
2	14	0	01
3	7	0	001
4	3	1	1001
5	1	1	11001
6	0	1	111001

(b) The algorithm converts N into a binary number. So, for example,

$$57 = (1 \times 32) + (1 \times 16) + (1 \times 8) + (0 \times 4) + (0 \times 2) + (1 \times 1) = 111001_2.$$

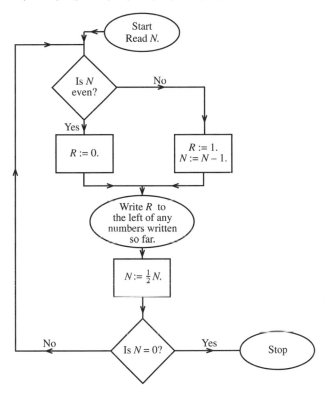

Sometimes when you see algorithms, you will see explanatory comments written about some or all of the steps. These will usually be put into curly brackets, { }, often called braces.

Here is an example of an algorithm with comments. It produces a permutation of the numbers 1 to n, that is, it produces the numbers 1 to n in random order, with no repeats. Notice how the comments make the algorithm easier to understand.

Step 1   Read n.   { n is the number of numbers in the permutation.}

Step 2 $i := 1$.   { i is the number of the permutation currently being found.}

Step 3 $r := Rand(1, n)$ { $Rand(1, n)$ is a random integer
                                                       between 1 and n inclusive.}

Step 4   If r has not already been used, write r, otherwise go to Step 3.
                                            { r is the ith number in the permutation.}

Step 5 $i := i + 1$.   {To get the next number in the permutation.}

Step 6   If $i < n$ go to Step 3, otherwise stop.

This algorithm is not very efficient, because you can spend much time regenerating random numbers which you have already found. You will find a much better algorithm in Miscellaneous exercise 1 Question 10.

Miscellaneous exercise 1

1 (a) Carry out the algorithm in this
        flow diagram for $x = 2$, $x = 3$
        and $x = 5$.

   (b) For what is this algorithm
        designed?

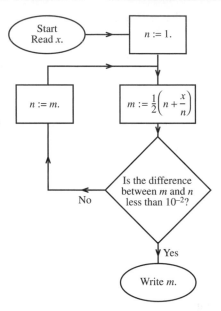

2 A flow diagram is shown below.

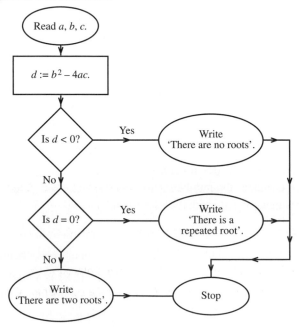

   (a) What is the output of this algorithm in the cases
        (i) $a = 1$, $b = 3$, $c = 2$;   (ii) $a = 1$, $b = 2$, $c = 1$;   (iii) $a = 1$, $b = 2$, $c = 3$?
   (b) What has this algorithm been designed to do?
   (c) How could you adapt the flow diagram to represent an algorithm to find the roots of a
        quadratic equation?

3 Euclid's algorithm is defined as follows.

Step 1 Read X and Y.

Step 2 If $X = Y$ then print X and stop.

Step 3 Replace the larger of X and Y by the difference between X and Y.

Step 4 Go back to Step 2.

(a) Carry out Euclid's algorithm for inputs of

   (i) 6 and 15,   (ii) 4 and 18,   (iii) 3 and 7.

(b) What does Euclid's algorithm find?

(c) Draw a flow diagram for this algorithm.

4 The Russian peasant's algorithm is defined as shown in the diagram.

Complete a table showing the successive values of x, y and t, when x and y initially take the values 11 and 9. Hence decide what the algorithm is designed to do.

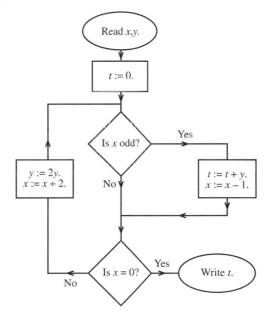

5 An algorithm to sort a list, L, of numbers into increasing order is defined as follows.

Step 1 Write down the first number of L as the start of a new list N.

Step 2 Take the next number of L. Compare it to each number of N in turn (from the left) and insert it at the appropriate place in N.

Step 3 Repeat Step 2 until all numbers of L have been placed in N.

Note that for each number from L, once you have found a number in N that is bigger than it, you do not need to make any further comparisons.

(a) Use this algorithm to sort $L = \{8, 12, 2, 54, 23, 31\}$. Write down the list N and count the number of comparisons at each stage.

(b) Order the list L in such a way that the algorithm will require the largest possible number of comparisons.

(c) What is the largest possible number of comparisons needed for this algorithm to sort a list of n numbers? What is the order of this algorithm?

6 Suppose there is a million-fold increase in a computer's speed. The computer can solve problems of linear order which are 1 million times as large as before, in the same amount of time. What is the equivalent effect on problems of

(a) quadratic order,   (b) order 2^n?

7 A plumber needs the following lengths, in centimetres, of copper piping for a job:

40, 40, 50, 60, 60, 70, 80, 100, 110, 120, 120, 130.

The piping comes in standard lengths of 2 m. How many lengths are needed if the plumber

(a) applies the First-Fit algorithm,

(b) applies the First-Fit Decreasing algorithm,

(c) is as efficient as possible?

8 The following algorithm is based on a method used by Archimedes to estimate π.

Step 1 $C := \frac{1}{2}\sqrt{3}$, $S := 3$, $T := 2\sqrt{3}$, $D := 2\sqrt{3} - 3$.

Step 2 Repeat

$$C := \sqrt{\tfrac{1}{2}(1+C)}.$$

$$S := \frac{S}{C}.$$

$$T := \frac{S}{C}.$$

$$D := T - S.$$

Until $D < 0.01$.

Step 3 Print S.

(a) Copy and complete the following table to show the values of C, S, T and D (rounded to 4 decimal places) for the first three runs through the repeat loop.

	C	S	T	D
Initial values	0.8660	3.0000	3.4641	0.4641
1st iteration	0.9659	3.1058	3.2154	0.1096
2nd iteration				
3rd iteration				

(b) Write down the number of runs through the repeat loop that are needed if the value 0.01 (in line 7 of the algorithm) is replaced by 0.1.

(c) Let d_n be the value of D from the nth iteration. Use the values of d_1, d_2 and d_3 to conjecture an approximate value for $\dfrac{d_{n+1}}{d_n}$.

(d) Hence suggest an approximate expression for the value of D from the nth iteration.

(OCR, adapted)

9 The Binary Search algorithm described below locates the position of a certain value, X, within an ordered sequence $S(1)$, $S(2)$, ... , $S(N)$.

$\text{INT}(0.5(I+J))$ gives the largest integer that is less than or equal to $0.5(I+J)$.

Step 1 $I := 1$, $J := N$. { I and J mark the boundaries of the sequence being searched.}

Step 2 If $I > J$ then print 'FAIL' and stop.   { X has not been found.}

Step 3 $M := \text{INT}(0.5(I+J))$.

Step 4 If $X = S(M)$ then print M and stop.   { X has been found.}

Step 5 If $X < S(M)$ then $J := M - 1$, otherwise $I := M + 1$.   {Reset boundaries.}

Step 6 Go to Step 2.

(a) Demonstrate carefully each step of the algorithm when it is applied to the sequence

     1   1   2   3   5   8   13   21

     (i)   with $X = 13$;

     (ii) with $X = 15$.

If N is a power of 2, the length of the sequence to be searched is at least halved at each iteration.

(b) If $N = 10$, work out the maximum possible length of the sequence to be searched at each iteration.

The efficiency of the algorithm can be measured by counting the number of runs through the repeat loop (in the worst case) for different values of N.

(c) Work out the efficiency for

     (i) $N = 8$,   (ii) $N = 16$.

10* This algorithm produces a random permutation of the numbers from 1 to n; that is, it produces the numbers from 1 to n in random order, with no repeats.

Step 1 Read n.   { n is the number of numbers in the permutation.}

Step 2 For $i = 1$ to n, $Perm(i) := i$.   { $Perm(i)$ will be the ith number in the permutation. This is an initialisation step.}

Step 3 $j := 1$.

Step 4 Repeat the following.

     $r := Rand(j, n)$.   { $Rand(j, n)$ is a random number from j to n inclusive.}

     Swap $Perm(j)$ with $Perm(r)$,

     $j := j + 1$.

     Until $j = n - 1$.

Step 5 Write $Perm(1)$, $Perm(2)$, ... , $Perm(n)$.

(a) Work through the algorithm with $n = 5$, in the case when the random numbers produced are successively, 3, 4, 3 and 5. What permutation do you finish with?

(b) The efficiency of this algorithm is the number of times that you have to use the random number generator. Find the efficiency for the case when there are n numbers.

(c) Compare this with the algorithm on page 11.

2 Graphs and networks

This chapter looks at problems which can be represented using graphs and networks. When you have completed it you should

- know what is meant by the terms 'arc', 'node', 'trail', 'path', 'tree', 'cycle', 'directed' and 'planar', as applied to graphs
- be able to use the orders of the nodes of a graph to determine if the graph is Eulerian or semi-Eulerian
- be familiar with some special graphs, namely complete graphs, bipartite graphs, trees and digraphs
- know Euler's relation on planar graphs, and be able to apply it.

2.1 Graphs and networks

Fig. 2.1 LRT Registered User No. 01/3566.

The standard London Underground map, the central part of which is shown in Fig. 2.1, shows how the various stations are connected. It does not attempt to show other properties, such as distances, or whether or not the track is above or below ground.

A simple way to model connectedness is with a **graph**, which consists of **nodes** (or **vertices**) joined by **arcs** (or **edges**). The graph shown in Fig. 2.2 has five nodes (the blobs) joined by four arcs.

Fig. 2.2

Note this quite different use of the word 'graph' from the conventional one.

The graph in Fig. 2.2 might represent one underground station which is linked directly to four other stations. It could just as easily represent the methane molecule CH_4, or a web site with five pages, one of which has links to the other four. In each case the graph serves as a model highlighting the connectedness of the original real-world situation.

Graphs are allowed to have **loops**, connecting nodes to themselves, as in Fig. 2.3a. They may also have **multiple arcs** between pairs of nodes, as in Fig. 2.3b.

Fig. 2.3a   Fig. 2.3b

However, in many contexts loops and multiple arcs are not appropriate. A graph without loops and multiple arcs is called a **simple** graph.

Fig. 2.4 shows all the simple graphs with 1, 2 and 3 nodes. There is one with 1 node, two with 2 nodes and eight with 3 nodes.

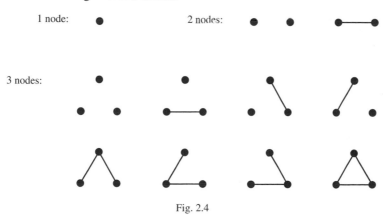

Fig. 2.4

All eleven of the graphs in Fig. 2.4 can be thought of as being contained in the final one, with 3 nodes and 3 arcs. In graph-theory terms you can say that each of them is a **subgraph** of the graph shown in Fig. 2.5.

Fig. 2.5

A graph in which each of the nodes is
connected by precisely one arc to every other
node is called a **complete graph**. The notation
K_n is used for the complete graph with n
nodes. Fig. 2.6 shows K_2, K_3 and K_4.

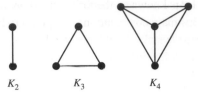

K_2 K_3 K_4

Each simple graph is a subgraph of K_n for any
sufficiently large n.

Fig. 2.6

Example 2.1.1

Find a formula for the number of arcs in the complete graph K_n.

Each node is connected to the other $n-1$ nodes and so it is at the end of $n-1$
arcs. There are n nodes and so there are $n \times (n-1)$ ends of arcs. As every arc has
two ends, there are $\frac{1}{2}n(n-1)$ arcs in total.

As a check, you can see that K_3 has $\frac{1}{2} \times 3 \times 2 = 3$ arcs, and K_4 has $\frac{1}{2} \times 4 \times 3 = 6$ arcs.

In addition to complete graphs, there is another important family of graphs, called
bipartite graphs. Bipartite graphs have two sets of nodes. The arcs only connect nodes
from one set to the other, and do not connect nodes within a set. Fig. 2.7 shows a
bipartite graph, with a set of 2 nodes in one oval, and a set of 3 nodes in another oval.

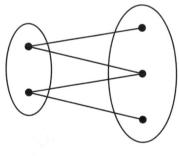

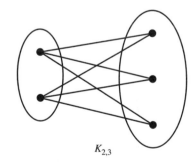

$K_{2,3}$

Fig. 2.7

Fig. 2.8

Note that it is not necessary for every node in one set to be connected to every node in
the other.

If, in a bipartite graph, every node in one set is connected to every node in the other set, the
graph is called a **complete bipartite graph**. If there are r nodes in one set, and s nodes in
the other, the complete bipartite graph is denoted by $K_{r,s}$. Fig. 2.8 shows $K_{2,3}$.

2.2   Leonhard Euler

Leonhard Euler was an 18th-century Swiss mathematician who made major
contributions to an enormous range of aspects of pure and applied mathematics, physics,
and astronomy. Much of the notation he used, including the symbols e and π, has
remained in standard use to the present day.

Euler is known as the father of graph theory. One
of his main contributions is reputed to have arisen
from a puzzle about the seven bridges over the
river Pregel in the Prussian city of Königsberg,
shown in Fig. 2.9. Can you find a 'circular' tour
which crosses each bridge precisely once?

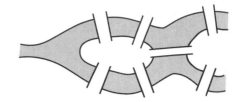

Fig. 2.9

A suitable graphical representation of the
Königsberg Bridge problem has the four land-masses
represented by nodes and the bridges represented by
arcs, as in Fig. 2.10.

Euler realised that, in any circular tour, a land-mass is
entered (via a bridge) the same number of times as it is
left (via a bridge). For a circular tour to exist, each
land-mass would therefore need to be linked to the
other land-masses by an even number of bridges. So a

Fig. 2.10

circular tour would require each node of Fig. 2.10 to have an even number of arcs coming
out of it. In fact, the numbers of arcs are 3, 3, 3 and 5, and so a circular tour is impossible.

A few definitions of graph theory terms are needed to move forward. These definitions
will be illustrated with reference to the section of the Underground map shown in
Fig. 2.11. You should regard *each* station as a node, not just the interchange stations.

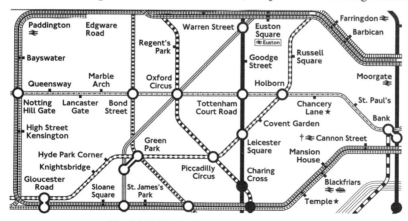

Fig. 2.11 LRT Registered User No. 01/3566.

A **trail** is a sequence of arcs such that the end node of one arc is the start node of the
next. One example is Green Park–Piccadilly Circus–Leicester Square–Charing Cross–
Piccadilly Circus–Oxford Circus–Bond Street. The term **route** is also used to mean trail.

A **path** is a trail with the restriction that no node is passed more than once. The trail
given above, which starts at Green Park, is therefore not a path, because Piccadilly
Circus occurs twice.

A **closed** trail is one where the initial and final nodes are the same. One example is Green Park–Piccadilly Circus–Leicester Square–Charing Cross–Piccadilly Circus–Green Park.

A **cycle** is a closed trail where only the initial and final nodes are the same. The closed trail given above is not a cycle itself but it contains the cycle, Piccadilly Circus–Leicester Square–Charing Cross–Piccadilly Circus. Although one node is allowed to be used twice, at the beginning and the end, no arc is allowed to be used twice.

The **order** of a node is the number of arcs meeting at that node. For example, Oxford Circus has order 6, Temple has order 2 and Holborn has order 4. If the number of arcs is even, the node has **even order**; if the number of arcs is odd, the node has **odd order**.

A **connected** graph is one where, for any two nodes, a path can be found connecting the two nodes.

2.3   Eulerian graphs

An **Eulerian graph** is a connected graph which has a closed trail containing every arc precisely once. Fig. 2.12 contains one Eulerian graph, and one graph which is not Eulerian.

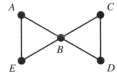

This has a closed trail
A–B–C–D–B–E–A
and is therefore Eulerian.

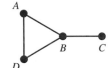

A closed trail containing
every arc would have to
contain arc *BC* twice. This
graph is therefore not Eulerian.

Fig. 2.12

It is a relatively easy task (see Exercise 2A Question 2) to prove that every node of an Eulerian graph must have even order.

Euler proved both this result and its converse (which is much more difficult to prove).

A connected graph is Eulerian if and only if every node has even order.

Example 2.3.1
(a) Explain why the graph in Fig. 2.13 is not Eulerian.

(b) Which single arc could be deleted in order to make the resultant graph Eulerian? Explain your answer and find a closed trail containing all the remaining arcs.

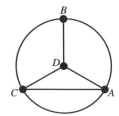

Fig. 2.13

(a) Nodes *B* and *D* have odd order.

(b) Removing arc *BD* creates a connected graph with every node having even order. This is therefore Eulerian. An example of a closed trail is *ABCACDA*.

A well known modern puzzle is to trace the
diagram shown in Fig. 2.14 without lifting the
pen off the paper and without going over any
lines twice.

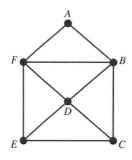

In graph theory terms the problem is therefore
to find a trail (not necessarily closed) which
contains every arc precisely once. The initial
and final nodes, if different, will then have odd
order, whereas all other nodes will have even
order.

Fig. 2.14

In Fig. 2.14, only nodes *C* and *E* have odd order and therefore the trick is to make sure you
start tracing from one or other of these two points. An example, is *CBAFBDCEDFE*.

A graph is called **semi-Eulerian** if it has a trail which is not closed that contains every
arc precisely once.

A connected graph is semi-Eulerian if and
only if precisely two nodes have odd order.

Exercise 2A

1 Which of the following graphs are Eulerian and which are semi-Eulerian?

(a)   (b)   (c)   (d)

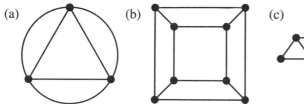
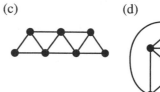
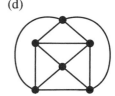

2 Explain carefully why an Eulerian graph can only have nodes of even order.

3 Draw connected graphs which have

   (a) 1 node of order 1 and 3 nodes of order 3,

   (b) 3 nodes of order 1 and 1 node of order 3.

4 Is it possible to find a route which passes
through every door precisely once?

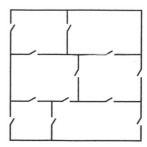

5   Think of one or two practical situations which it would be appropriate to model with a graph with multiple edges.

6   (a)   What is the order of a node of K_n?

      (b)   For what value of n is K_n Eulerian or semi-Eulerian?

7   (a)   How many arcs are there in the complete bipartite graph $K_{r,s}$?

      (b)   Analyse $K_{r,s}$ in terms of being Eulerian or semi-Eulerian.

8   Four Members of Parliament, Ann, Brian, Clare and David, are being considered for four cabinet posts. Ann could be Foreign Secretary or Home Secretary, Brian could be Home Secretary or the Chancellor of the Exchequer, Clare could be Foreign Secretary or the Minister for Education and David could be the Chancellor or the Home Secretary.

      (a)   Draw a bipartite graph to represent this situation.

      (b)   How many options does the Prime Minister have?

2.4   Planar graphs

Most of the graphs you have seen so far could have been drawn on a flat surface without arcs crossing one another. However, the Underground map contains lines which do cross under and over each other.

A **planar graph** is one which can be drawn in a plane in such a way that arcs only meet at nodes.

For example, Fig. 2.15a shows a planar graph, because it can be drawn as shown in Fig. 2.15b.

Fig. 2.15a

Fig. 2.15b

For planar graphs which are actually drawn in a plane, such as Fig. 2.15b, Leonhard Euler noticed an important relationship between the numbers of arcs, nodes and regions (the areas the plane is divided up into by the graph). Consider a few examples of such drawings.

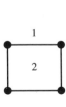

Fig. 2.16a

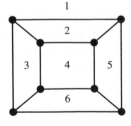

Fig. 2.16b

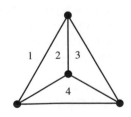

Fig. 2.16c

You can see that the graph in Fig. 2.16a has 4 arcs, 4 nodes and 2 regions, the graph in Fig. 2.16b has 12 arcs, 8 nodes and 6 regions, and the graph in Fig. 2.16c has 6 arcs, 4 nodes and 4 regions.

What Euler noticed, and then proved, was the following.

> **Euler's relationship**   For any connected graph drawn
> in the plane with R regions, N nodes and A arcs,
>
> $$R + N = A + 2.$$

Note that this relationship is true even for a single node, which has $A = 0$, $N = 1$ and
$R = 1$. Throughout this section the letters R, N and A will be used to denote the
number of regions, nodes and arcs of the graph being studied.

Euler's relationship can be useful when deciding whether or not a
graph is planar. For example, consider the well known puzzle
about three houses and three utilities, where you have to connect
each house to each utility without a cross-over. See Fig. 2.17.

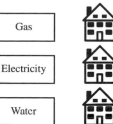

This puzzle turns out to be impossible, apart from 'trick'
solutions such as running a supply line through a house or power
station. However, it is surprisingly difficult to prove that the
puzzle is impossible to solve.

Fig. 2.17

If you reduce the problem to a problem in graph theory, you
obtain the graph in Fig. 2.18.

This consists of two sets of three points, and arcs joining every
one of one set of three points to every one of the other set of three
points. It is therefore the complete bipartite graph, $K_{3,3}$.

Fig. 2.18

Example 2.4.1
Show that the bipartite graph $K_{3,3}$ is not a planar graph.

Suppose $K_{3,3}$ is planar. Then you can apply Euler's relationship.

For $K_{3,3}$, $A = 9$ and $N = 6$, so $R + 6 = 9 + 2$, giving $R = 5$.

Now consider one of these regions. Its boundary forms a cycle.

Alternate points around such a cycle must be in first one and then the other of the
sets of three points. The cycle therefore has an even number of points. A cycle of
length two is not possible because there is at most one arc between any two points.
Therefore each region has at least 4 boundary arcs.

So, as there are 5 regions, counting arcs round all boundaries gives at least 5×4
arcs. Each arc separates 2 regions and so the graph must have at least
$\frac{1}{2} \times 5 \times 4 = 10$ different arcs. But there are only 9 arcs, so there is a contradiction.
It follows that $K_{3,3}$ cannot be planar.

The complete graphs K_n and $K_{m,n}$ turn out to be important in the theory of planar graphs.
There is an important (and difficult to prove!) result called Kuratowski's theorem which says
that a graph is non-planar if and only if it involves at least one of $K_{3,3}$ or K_5.

2.5   Trees

A tree is a connected graph with no cycles. All of the graphs in Fig. 2.19 are trees.

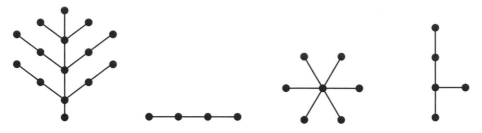

Fig. 2.19

Trees were important in the early history of graph theory because they were used by the 19th-century English mathematician, Sir Arthur Cayley, to study the possible structure of saturated hydrocarbons. Fig. 2.20,

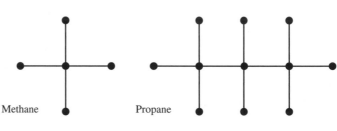

Methane   Propane

Fig. 2.20

in which nodes of order 4 represent carbon atoms and nodes of order 1 represent hydrogen atoms, shows methane, CH_4, and propane, C_3H_8.

Trees are also important in the study of connected graphs. Any connected graph contains at least one subgraph which is a tree connecting every node of the original graph. The complete graph K_4 has many such trees. Fig. 2.21 shows just one example obtained by removing successive arcs.

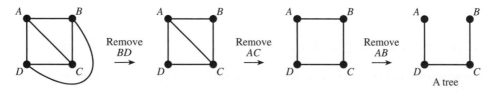

Fig. 2.21

Example 2.5.1

The number of nodes in a connected planar graph is one greater than the number of arcs. Prove that the graph is a tree.

According to Euler's relationship, $R + N = A + 2$.

You are given that $N = A + 1$, so, substituting this in $R + N = A + 2$ gives $R = 1$.

The graph has only one region and therefore no cycles, so it is a tree.

Exercise 2B

1 By redrawing the graph in the diagram show that it is planar.

2 Check Euler's relationship for each of the graphs below.

   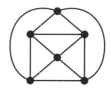

3 Draw all the essentially different trees with 3, 4, 5 and 6 nodes.

4 Give an example of
(a) a planar graph with every node of order 4,
(b) a planar graph with every region bounded by 5 arcs.

5 Show that K_4 is planar.

6 Show that $K_{2,n}$ is planar for any value of n.

7 Every region of a planar drawing of a graph has three boundary arcs.
(a) Explain why $3R = 2A$.
(b) Use Euler's relation to show that $A = 3N - 6$.
(c) Draw such a graph with $N = 5$.

8* Prove that K_5 is not planar. (Note that if K_5 were planar then any region of the graph would have at least 3 boundary arcs and therefore $2A \geqslant 3R$.)

2.6 Network problems

For many problems, you need to know more than just whether nodes are connected or not. For example, problems involving the distance between towns or the costs of various links require a numerical value to be given to each arc. Such a numerical value is given the general title of a **weight**. A graph whose arcs have weights can be called either a 'weighted graph' or a **network**.

A simple network is shown in Fig. 2.22.

Most of the remainder of this book is concerned with problems about networks. You might like to try to think of as many different types of questions as you can about the network in Fig. 2.22.

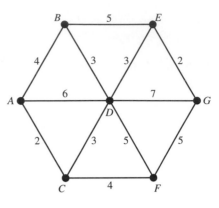

Here are some possibilities.

Fig. 2.22

- The connections of minimum weight which connect all of the nodes. This is called the minimum connector problem; see Chapter 3.
- The route of minimum weight from A to G. This is the shortest path problem; see Chapter 4.
- The closed trail of minimum weight which includes every arc at least once. This is the route inspection problem; see Chapter 5.
- The cycle of minimum weight which includes every node. This is the travelling salesperson problem; see Chapter 6.

2.7   Directed graphs

Sometimes the links which are represented by arcs have a direction associated with them. For example, some roads are one-way, and a page on one company's web site may have a link to that of an advertiser but without there being any reciprocal link.

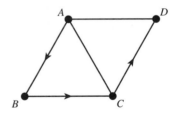

Fig. 2.23

A graph with directed arcs is known as a directed graph, or **digraph**.

In Fig. 2.23, the digraph has three directed arcs, and two arcs which can be traced in either direction.

Example 2.7.1
(a) Show that the digraph of Fig. 2.24 is Eulerian by finding all the closed trails which contain every directed arc.

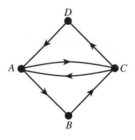

Fig. 2.24

(b) At each node of this Eulerian digraph, what is true about the number of arcs in and the number of arcs out?

(a) *ABCDACA, ABCACDA*

(b) At A, number of arcs in = number of arcs out = 2.
At B, number of arcs in = number of arcs out = 1.
At C, number of arcs in = number of arcs out = 2.
At D, number of arcs in = number of arcs out = 1.

The number of arcs in is equal to the number of arcs out, at every node.

2.8 Representing networks by matrices

It is important to have a way of representing networks which does not rely on a diagram. In particular, computers cannot work with diagrams, so how can you tell a computer all the essential detail about a network?

The answer is surprisingly simple. A table showing all the information about the weights in a network is all that you need. A table of this type is called a **matrix**.

For example, you can summarise the information in Fig. 2.25 in the matrix

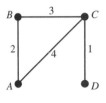

$$\begin{array}{c c c c c} & A & B & C & D \\ A & \begin{pmatrix} - \\ 2 \\ 4 \\ - \end{pmatrix} & \begin{matrix} 2 \\ - \\ 3 \\ - \end{matrix} & \begin{matrix} 4 \\ 3 \\ - \\ 1 \end{matrix} & \begin{matrix} - \\ - \\ 1 \\ - \end{matrix} \end{array}$$

Fig. 2.25

Notice that for a network in which all the links are two-way links, the matrix is symmetrical about the diagonal line drawn from the top left to the bottom right of the matrix. This diagonal is called the main diagonal of the matrix.

You can also store the information about weights in a digraph using a similar method, but the resulting matrix is no longer symmetrical about the main diagonal.

For example, you can summarise the information in Fig. 2.26 in the matrix

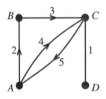

$$\begin{array}{c} \text{To} \\ \begin{array}{c c c c c} & A & B & C & D \\ A & \begin{pmatrix} - \\ - \\ 5 \\ - \end{pmatrix} & \begin{matrix} 2 \\ - \\ - \\ - \end{matrix} & \begin{matrix} 4 \\ 3 \\ - \\ 1 \end{matrix} & \begin{matrix} - \\ - \\ 1 \\ - \end{matrix} \end{array} \end{array}$$

From

Fig. 2.26

As the matrix is no longer symmetrical it is important to make the directions of the links clear. If there is ambiguity the words 'From' and 'To' will be inserted to make it clear, for example, that the entry 5 means that there is an arc of weight 5 from C to A.

If you reconstruct the network from the matrix, you should realise that your reconstruction may appear very different from someone else's reconstruction. For example, the digraph in Fig. 2.27 gives rise to the same matrix as the digraph in Fig. 2.26.

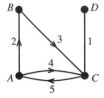

Fig. 2.27

Exercise 2C

1 (a) A network is used to represent the times taken to travel between various cities. Why might directed arcs be appropriate?

   (b) What type of electrical networks are best represented by

   (i) digraphs,   (ii) undirected graphs?

2 Write down two matrices which represent the graph and the digraph shown in the figure.

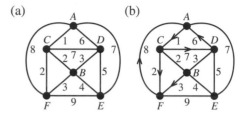

3 For the network shown in the diagram,

   (a) find the connections of minimum weight which connect all the nodes;

   (b) find the route from A to G of minimum weight.

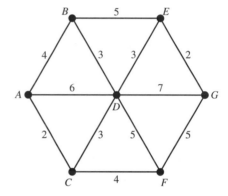

4 Draw several trees, with nodes of orders 1 or 4 only, to represent possible saturated hydrocarbons. What is the connection between the number of hydrogen atoms (nodes of order 1) and the number of carbon atoms (nodes of order 4)?

5 Design a flow diagram for an algorithm that deletes arcs from any connected graph and produces a tree on the same nodes. (You must ensure that deleting an arc does not disconnect the graph.)

6 The 'even-weight code of length 3' uses four code-words, 000, 011, 101 and 110. The 'distance' between two of these code-words is the number of places in which the binary digits differ.

   (a) Draw a network to illustrate the distances between the four code-words.

   (b) When transmitting and receiving codes, what is the advantage of using code-words in which all distances are at least 2?

7 Suppose that at a social gathering the number of handshakes that occur is H. Suppose further that there are n people, who shake hands h_1, h_2, … and h_n times respectively.

   (a) What, in terms of H, is the sum of all the h_i?

   (b) Represent people by nodes and handshakes by arcs. What relationship does the result in part (a) imply about the orders of the nodes and the numbers of arcs?

Miscellaneous exercise 2

1 (a) Count the numbers of faces, vertices and edges for a cube, a square-based pyramid and a tetrahedron, and put the results into a table.

(b) What relationship can you see for the numbers in your table? Give an explanation for the relationship.

2 (a) For the system of islands and bridges shown in the diagram, is it possible to find a trail which crosses every bridge precisely once? Carefully explain your reasoning.

(b) Is it possible to find a closed trail with the same property?

3 An algorithm for deciding whether a graph is planar or not is described below. (Note that it does not work for all graphs.)

Step 1 Find a cycle which passes through every node of the graph. If no such cycle can be found then report this fact and stop.

Step 2 Redraw the graph with this cycle drawn as a circle and with all arcs not included in the cycle drawn as chords of the circle.

Step 3 Draw a new graph whose nodes represent the chords of the circle. Join two of these new nodes with an arc if the relevant chords cross each other.

Step 4 If the new graph is *not* bipartite then the original graph is non-planar. If the new graph *is* bipartite then the original graph is planar. Use the natural subdivision of the nodes of the bipartite graph to split the chords of Step 2 into two sets. Choose one of these sets of chords and redraw them as arcs *outside* the circle.

Demonstrate the use of this algorithm by applying it to each of the following graphs. In each case show sufficient detail to make your working clear.

(a)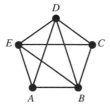

(b)

(c)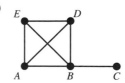

3 Minimum connector problems

This chapter looks at effective ways of connecting nodes of a network. When you have completed it you should

- know what a spanning tree for a connected graph is
- know the term 'greedy algorithm', and the steps of the greedy algorithms called Prim's and Kruskal's algorithms
- be able to find a spanning tree of minimum weight by using either Prim's algorithm or Kruskal's algorithm
- know that Prim's algorithm can easily be programmed for a computer, and be able to apply the algorithm in matrix form.

3.1 Introduction

To create an internal computer network in a school, cabling has to be laid between the five main computer areas. These five areas and the costs of the various alternative runs of cables are as shown in Fig. 3.1.

To connect all of the areas together just four runs of cable are needed. Three of the many possibilities are shown in Fig. 3.2.

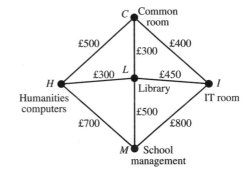

Fig. 3.1

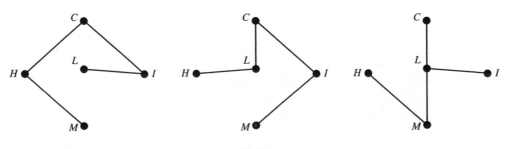

Fig. 3.2

Find the costs of each one of these three possibilities. Can you find a cheaper way of laying the cable? (See Exercise 3A Question 3.)

You may well have been able to spot the solution to the simple problem posed above. However, for problems such as connecting cable TV to all the main areas of a town, linking up houses to the national electricity grid, or joining up soldering points on a printed circuit, the number of possible connections is so large that you have to use an algorithm that can be performed by a computer. The purpose of this chapter is to develop such an algorithm.

3.2 Spanning trees

You may have noticed that each of the possible runs of cable considered in Section 3.1 is a tree connecting together all of the nodes. Any tree which connects all the nodes of a graph is called a **spanning tree** for that graph.

Notice that each of the spanning trees contains the same number of arcs, and that this number is one less than the number of nodes of the graph. This is illustrated in Fig. 3.3.

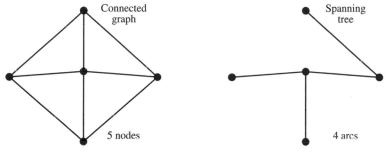

Fig. 3.3

This relationship between the number of nodes of the connected graph and the number of arcs of the spanning tree is always true and is proved in Section 3.6.

> For a connected graph with n nodes, each spanning tree has precisely $n-1$ arcs.

Example 3.2.1

Find the number of spanning trees of the graph in Fig. 3.4.

The graph has three nodes, so each spanning tree has two arcs. There are three different ways of deleting an arc from the original graph, so the spanning trees are the three shown in Fig. 3.5.

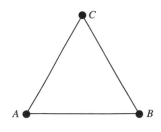

Fig. 3.4

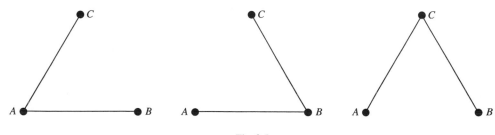

Fig. 3.5

Remember that a 'complete' graph is one which has every possible arc. The number of possible spanning trees for a complete graph increases very rapidly as the number of nodes increases (see Table 3.6).

Number of nodes	Number of spanning trees
3	3
4	16
...	...
20	2.6×10^{23}
...	...

Table 3.6

Fig. 3.7 shows the complete graph with four nodes. Fig. 3.8 shows the 16 possible spanning trees.

Fig. 3.7

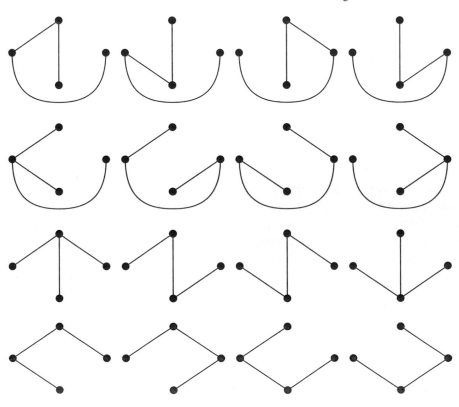

Fig. 3.8

For graphs with a reasonably large number of nodes, it is not possible for even the most powerful computer to scan all the possible spanning trees to find the best one for a problem like those in Section 3.1. To do it in a reasonable amount of time, you need an algorithm.

3.3   Prim's algorithm

The spanning tree of minimum weight is called the **minimum spanning tree**, or the **minimum connector**. For a given connected graph, Prim's algorithm is a quick method of finding the minimum spanning tree. The sequence of steps to be followed is shown below.

> **Prim's algorithm**   To find a minimum spanning tree T:
>
> **Step 1**   Select any node to be the first node of T.
>
> **Step 2**   Consider the arcs which connect nodes in T to nodes outside T. Pick the one with minimum weight. Add this arc and the extra node to T. (If there are two or more arcs of minimum weight, choose any one of them.)
>
> **Step 3**   Repeat Step 2 until T contains every node of the graph.

Example 3.3.1

Use Prim's algorithm to obtain a minimum spanning tree for the graph in Fig. 3.9.

The successive stages, starting with node C, are as shown in Fig. 3.10.

Notice that at the fourth stage, either FA or FB could have been chosen.

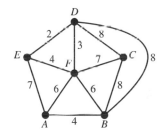

Fig. 3.9

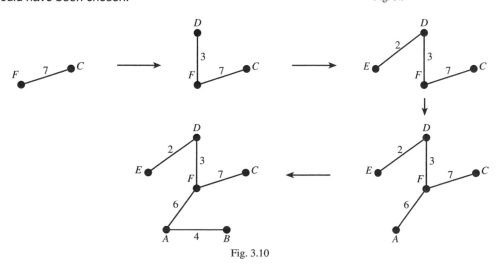

Fig. 3.10

The minimum spanning tree has weight $7 + 3 + 2 + 6 + 4 = 22$ units.

Notice that when you apply Prim's algorithm, you are simply choosing the arc which is immediately 'best' without being concerned about the long-term consequences of your choice. The fact that this 'greedy' approach to the minimum connector problem always leads to the best solution is justified in Section 3.6.

3.4   Kruskal's algorithm

One of several alternative algorithms for finding the minimum spanning tree uses arc weights directly rather than considering connecting up points. It was invented by an American mathematician, Martin Kruskal.

> **Kruskal's algorithm**
>
> To find a minimum spanning tree for a connected graph with n nodes:
>
> **Step 1**   Choose the arc of least weight.
>
> **Step 2**   Choose from those arcs remaining the arc of least weight which does *not* form a cycle with already chosen arcs. (If there are several such arcs, choose one arbitrarily.)
>
> **Step 3**   Repeat Step 2 until $n-1$ arcs have been chosen.

Example 3.4.1

Apply Kruskal's algorithm to the network in Fig. 3.11.

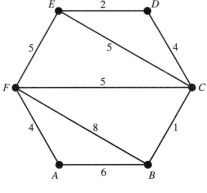

Fig. 3.11

Starting with the arc of least weight, which is BC, of weight 1, Kruskal's algorithm gives the results in Table 3.12.

Arc	Weight	Choice
BC	1	1st
DE	2	2nd
AF	4	3rd
CD	4	4th
CE	5	Not chosen
CF	5	5th
EF	5	–
AB	6	–
BF	8	–

Table 3.12

The resulting minimum spanning tree is shown in Fig. 3.13.

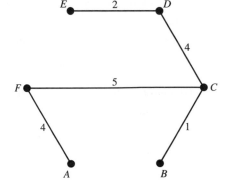

Fig. 3.13

Kruskal's algorithm is very easy to apply to a small network. The need to spot whether or not a cycle has been produced makes it less good for more involved questions and also less easy to program for a computer.

Exercise 3A

1 Find all the possible spanning trees for the graph in the figure.

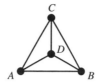

2 (a) Draw all possible spanning trees for this network.

    (b) Which of the spanning trees has minimum weight?

    (c) Use Prim's algorithm, starting with node A, to find the minimum spanning tree.

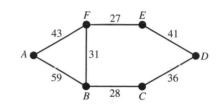

3 Find the cheapest way of laying the cable for the problem posed at the beginning of Section 3.1, whose diagram is reproduced here.

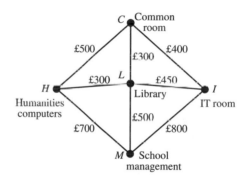

4 Consider the network of Question 2. Show that Kruskal's algorithm produces the same minimum spanning tree as Prim's algorithm.

5 Apply Kruskal's algorithm to the network shown here. Draw a diagram showing the minimum connector, state the order in which you added arcs and work out the total length of the minimum connector.

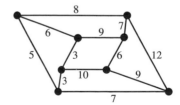

6 To obtain a minimum spanning tree, you can *delete* arcs in order of decreasing weight.

    (a) When should an arc not be deleted?

    (b) Devise an algorithm for obtaining a minimum spanning tree by such a method of deletion.

3.5 Matrix formulation

Prim's algorithm can be translated relatively easily into a computer program. The best starting point for this is the matrix which shows the weights on the various arcs.

The graph used in Example 3.3.1 is tabulated in Table 3.14.

	A	B	C	D	E	F
A	–	4	–	–	7	6
B	4	–	8	8	–	6
C	–	8	–	8	–	7
D	–	8	8	–	2	3
E	7	–	–	2	–	4
F	6	6	7	3	4	–

Table 3.14

The solution to Example 3.3.1 would then proceed as follows.

Select node C to be the first node of T. Circle C in the top row to show that you have selected it, and cross out the C row.

Look for the smallest weight in the columns of the nodes in T (that is, the C column) and circle it. This is a 7, in row F. So F becomes the second node of the spanning tree. Select F by circling it in the top row, and cross out the F row. This gives Table 3.15.

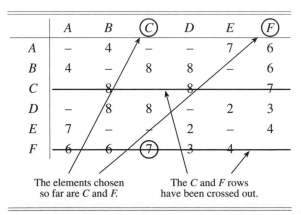

The elements chosen so far are C and F.
The C and F rows have been crossed out.

Table 3.15

Now circle the smallest weight in the columns corresponding to the nodes of T (that is, the C and F columns). This weight is 3, in row D, so choose D to be the third node of T: circle the D in the top row, and cross out the D row, as shown in Table 3.16.

	A	B	Ⓒ	Ⓓ	E	Ⓕ
A	–	4	–	–	7	6
B	4	–	8	8	–	6
C	–	8	–	8	–	7
D	–	8	8	–	2	③
E	7	–	–	2	–	4
F	6	6	⑦	3	4	–

Table 3.16

You can continue this procedure, successively choosing E, A and B. The final state of the matrix is shown in Table 3.17. The circled numbers give you the minimum spanning tree as found in Example 3.3.1.

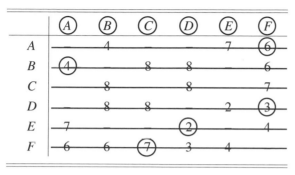

Table 3.17

The matrix formulation of Prim's algorithm to find a minimum spanning tree T is:

Step 1 Select any node to be the first node of T.

Step 2 Circle the new node of T in the top row, and cross out the row corresponding to this new node.

Step 3 Find the smallest weight left in the columns corresponding to the nodes of T, and circle this weight. Then choose the node whose row the weight is in to join T. (If there are several possibilities for the weight, choose any one of them.)

Step 4 Repeat Steps 2 and 3 until T contains every node.

Example 3.5.1

Find a minimum spanning tree for the weighted graph with the following matrix form.

	A	B	C	D	E	F	G	H
A	–	23	17	–	18	–	15	–
B	23	–	9	10	12	16	–	14
C	17	9	–	9	20	–	27	–
D	–	10	9	–	–	–	–	16
E	18	12	20	–	–	7	20	–
F	–	16	–	–	7	–	24	17
G	15	–	27	–	20	24	–	–
H	–	14	–	16	–	17	–	–

Select A as the first node of T. Circle A in the top row and cross out the A row.

Circle the smallest weight in the A column, which is 15. This corresponds to G, which becomes the next node of T. Circle G in the top row and cross out the G row.

The smallest remaining weight in the columns for nodes of T is 17, corresponding to C. Circle the 17, and C in the top row, and cross out the C row. At this stage the situation is:

	Ⓐ	B	Ⓒ	D	E	F	Ⓖ	H
A	—	~~23~~	~~17~~		~~18~~		~~15~~	
B	23	–	9	10	12	16	–	14
C	⑰	~~9~~		~~9~~	~~20~~		~~27~~	
D	–	10	9	–	–	–	–	16
E	18	12	20	–	–	7	20	–
F	–	16	–	–	7	–	24	17
G	⑮		~~27~~		~~20~~	~~24~~		
H	–	14	–	16	–	17	–	–

Either of the two 9s in the C column could now be chosen. Arbitrarily, choose D to be the next node of T. Circle the 9, and the D in the top row, and cross out the D row.

Continue this process. You can carry out all the working in a single table:

Order of selection		Ⓐ	Ⓑ	Ⓒ	Ⓓ	Ⓔ	Ⓕ	Ⓖ	Ⓗ
1	A	—	~~23~~	~~17~~		~~18~~		~~15~~	
5	B	~~23~~		⑨	~~10~~	~~12~~	~~16~~		~~14~~
3	C	⑰	~~9~~		~~9~~	~~20~~		~~27~~	
4	D	—	~~10~~	⑨					~~16~~
6	E	~~18~~	⑫	~~20~~			7	~~20~~	
7	F		~~16~~			⑦		~~24~~	~~17~~
2	G	⑮		~~27~~		~~20~~	~~24~~		
8	H		⑭		~~16~~		~~17~~		

The minimum spanning tree is illustrated in Fig. 3.18.

The minimum weight is
$15 + 17 + 9 + 9 + 12 + 14 + 7 = 83$
units.

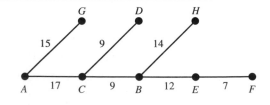

Fig. 3.18

Exercise 3B

1 The road distances in kilometres between eight towns are shown in the table. Boxes with thick borders refer to motorway routes.

	A	B	C	D	E	F	G	H
A	–	20	45	50	60	50	40	50
B	20	–	50	60	50	65	30	30
C	45	50	–	30	10	80	80	75
D	50	60	30	–	55	70	85	100
E	60	50	10	55	–	85	75	55
F	50	65	80	70	85	–	70	100
G	40	30	80	85	75	70	–	45
H	50	30	75	100	55	100	45	–

(a) Use Prim's algorithm, starting with A, to find the minimum connector for the towns. Draw a diagram showing the minimum connector, state the order in which you added arcs, and work out the total length of the minimum connector.

(b) The time for journeys can be estimated by assuming an average speed of 100 km h^{-1} for the motorway routes and 60 km h^{-1} for the other roads. Complete a table of times, in minutes, between the towns. Find the minimum connector for the times, and write down the total of all the times on the minimum connector.

(c) Explain why the answer for part (b) is not the total time for the minimum connector found in part (a).

2 The costs in £ sterling of tickets for direct flights between six cities are shown in the table.

	A	B	C	D	E	F
A	–	45	60	50	90	145
B	45	–	70	25	80	110
C	60	70	–	55	70	320
D	50	25	55	–	35	175
E	90	80	70	35	–	80
F	145	110	320	175	80	–

(a) Use Prim's algorithm to construct a minimum connector. Show all your working clearly and state the order in which you add arcs or nodes.

(b) Suppose that, in addition to the cost of tickets, an airport tax of 10% must be paid on leaving any airport. What effect will this have on the minimum connector?

(c) Suppose that airport taxes are actually £15 for any flight involving airports D, E or F and £5 otherwise. Work out the new minimum connector.

3 The figure shows the distances in kilometres along recommended motoring routes between ten French towns.

Dijon

296   Grenoble

462   714   Le Mans

507   282   926   Marseille

662   334   1081   188   Nice

297   549   138   761   916   Orléans

313   565   203   776   931   130   Paris

515   560   182   733   888   212   330   Poitiers

244   139   507   309   464   392   513   421   St-Etienne

415   534   82   745   900   112   234   100   425   Tours

(a) Use Prim's algorithm, starting with Paris, to find the minimum connector. Show all your working clearly, and work out the minimum connector's total length.

(b) The distances from Geneva (in Switzerland) to the ten French towns are as follows.

D	G	L	M	N	O	Pa	Po	S	T
199	144	634	434	477	496	537	501	164	528

What is the total length of the minimum connector when Geneva is included?

3.6* Justification of Prim's algorithm

The aim of this section is to show that, for any connected graph, Prim's algorithm in the form given in Section 3.3 will always produce a spanning tree of minimum possible weight. You may omit this section and assume the result if you wish.

Consider the formulation of Prim's algorithm given in Section 3.3.

> To find a minimum spanning tree T:
>
> **Step 1** Select any node to be the first node of T.
>
> **Step 2** Consider the arcs which connect nodes in T to nodes outside T. Pick the one with minimum weight. Add this arc and the extra node to T. (If there are two or more arcs of minimum weight, choose any one of them.)
>
> **Step 3** Repeat Step 2 until T contains every node of the graph.

In the justification, the following notation is used: T is formed by successively adding nodes $v_1, v_2, \ldots, v_n$ and arcs $a_1, a_2, \ldots, a_{n-1}$.

It might seem more natural to call the nodes n_1, n_2, ... But n usually means the number of nodes, so to avoid confusion the nodes are called v_1, v_2, ... (v being short for 'vertex', another word for 'node').

The justification requires three results along the way. The first task is to prove that Prim's algorithm does always produce a spanning tree.

Result 1   For a connected graph G with n nodes, Prim's algorithm produces a subgraph T which
(a) has $n-1$ arcs and n nodes,
(b) is connected,
(c) has no cycles,
(d) is a spanning tree.

Justification
(a) For Step 2 to be no longer possible, there would have to be no connections between nodes of T and any nodes not in T. Provided the original graph is connected, Step 2 can therefore always be repeated until T contains every node.

Therefore T has n nodes. Starting from the initial node of T, each application of Step 2 adds one arc and one node to T. The number of arcs of T is therefore one less than the number of nodes, which is $n-1$.

This completes the justification of (a).

Denote by T_i the graph formed by the addition of the node v_i.

(b) By Step 1, T_1 consists simply of the node v_1. By Step 2, node v_2 is joined to T_1 by arc a_1, so T_2 is connected.

Using Step 2 again, node v_3 is joined to T_2 by arc a_2. T_2 is connected, so there must be a route from v_3 to any node in T_2. So T_3 is also connected.

Repeat this argument until all the nodes have been added. The final stage shows that T_n is connected. But $T_n = T$, so T is connected.

(c) Suppose T has a cycle. Then let v_k be the node in this cycle with maximum possible k. In the cycle, v_k is joined by two arcs to other nodes. These nodes must be in T_{k-1}, because of the maximality of k. However, by the definition of Prim's algorithm, v_k is only joined by a single arc to nodes of T_{k-1}. This is a contradiction, so T has no cycles.

(d) From part (b), T is connected. It also has no cycles, by (c), and is therefore a tree. By (a) it spans G, because it has the same number of nodes as G, and it is therefore a spanning tree.

Result 1 shows that a spanning tree with n nodes produced by Prim's algorithm has $n-1$ arcs. In fact, any tree with n nodes has $n-1$ arcs; this is Result 2.

Result 2   Any tree with n nodes has precisely $n-1$ arcs.

Justification
Let S be any tree with n nodes. Then S is a connected graph, so, by Result 1, S must have a spanning subtree, T, with $n-1$ arcs.

If $S = T$, then there is nothing to prove. So suppose that S has at least one arc other than those in T, and suppose this arc connects node v to node w (see Fig. 3.19). But T is connected, so w is connected to v by a path of arcs in T. Then, in S, there is a cycle.

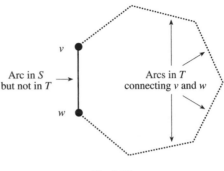

But S is a tree and so contains no cycles. Thus the supposition that S has at least one arc other than those in T leads to a contradiction. Thus the only possibility is $S = T$.

Fig. 3.19

Result 1 shows that, for any connected graph with n nodes, Prim's algorithm will produce a spanning tree. It only remains to show that it always finds a spanning tree of minimum weight.

In the justification of Result 3 you may find it helpful to refer to the example in Fig. 3.20. However, the proof that Prim's algorithm gives a tree of minimum weight is general, and does not rely on the diagrams.

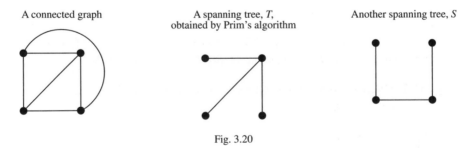

Fig. 3.20

For the example in Fig. 3.20, T_1, T_2, T_3 and T_4 are as shown in Fig. 3.21.

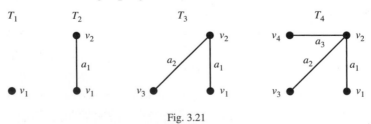

Fig. 3.21

Result 3   Let S be any spanning tree for a connected graph and let T be a spanning tree produced by Prim's algorithm. Then $\text{weight}(T) \leqslant \text{weight}(S)$.

Justification

The spanning tree S contains T_1. If S contains T_n, which is the same as T, then there is nothing to prove, so suppose that S contains T_k but not T_{k+1}. Thus a_k is not an arc of S.

Using the example, Fig. 3.22 shows the case when $k = 2$. Here S contains T_2 but not T_3.

Fig. 3.22

Add arc a_k to S. As S was a tree there is now a cycle containing the arc a_k. This arc a_k joins a node of T_k to a node outside T_k so there is at least one other arc of the cycle, f say, which joins a node of T_k to a node outside T_k. Remove arc f to form a new graph, S'.

Fig. 3.23 shows how this works with the example.

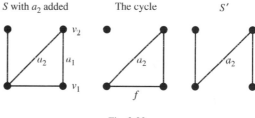

S with a_2 added   The cycle S'

Fig. 3.23

Arc f was part of a cycle so removing f does not disconnect the graph. S' is therefore a new spanning tree which contains T_{k+1}.

Since a_k was chosen to be the kth arc by Prim's algorithm, its weight is less than or equal to the weight of any other arc which joins a node of T_k to a node outside T_k. In particular,

$$\text{weight}(a_k) \leqslant \text{weight}(f),$$

and

$$\text{weight}(S') = \text{weight}(S) + \text{weight}(a_k) - \text{weight}(f) \leqslant \text{weight}(S).$$

By repeating this process as necessary, you can successively replace arcs of S by arcs of T, without increasing the weight. Eventually all $n-1$ arcs of S will have been replaced by the $n-1$ arcs of T and so the weight of T must be less than or equal to the weight of S. As this is true whatever spanning tree is used for starting the process, T is a minimum spanning tree.

This completes the justification that Prim's algorithm always produces a minimum spanning tree. Note that the minimum spanning tree need not be unique. There could be more than one, but they would all have the same weight.

Example 3.6.1

Look at Fig. 3.24. For the connected graph at the left, use the method of Result 3 to replace the arcs of the spanning tree S by those of spanning tree T, produced by Prim's algorithm.

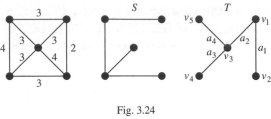

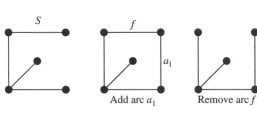

Fig. 3.24

The first arc in T which is not in S is a_1. Add a_1 to S. Then remove arc f from the cycle formed. This is shown in Fig. 3.25. Call the new spanning tree S'.

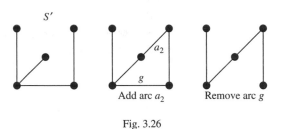

Add arc a_1 Remove arc f

Fig. 3.25

This increases the weight by 2, and then reduces it by 3.

The next arc in T which is not in the new spanning tree S' is a_2. Add a_2, and then remove arc g from the cycle formed. This is shown in Fig. 3.26. Call the new spanning tree S''.

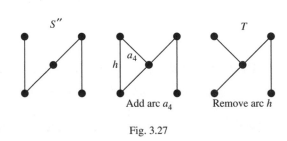

Add arc a_2 Remove arc g

Fig. 3.26

This increases the weight by 3, and reduces it by 3.

The next arc in T which is not in the new spanning tree S'' is a_4. (Note that a_3 is already in the new spanning tree.) Add a_4, and then remove arc h from the cycle formed. This is shown in Fig. 3.27.

Add arc a_4 Remove arc h

Fig. 3.27

This increases the weight by 3, and reduces it by 4. The net result of the process is to obtain T, with a weight of 2 less than S.

Try removing one of the other possible arcs in place of f in the first stage, and see what happens subsequently.

Miscellaneous exercise 3

1 The table shows the distances in miles between six US cities.

	C	Da	De	LA	NY	W
Chicago	–	800	900	1800	700	650
Dallas	800	–	650	1300	1350	1200
Denver	900	650	–	850	1650	1500
Los Angeles	1800	1300	850	–	2500	2350
New York	700	1350	1650	2500	–	200
Washington DC	650	1200	1500	2350	200	–

Use Prim's algorithm to find the minimum connector. Draw the minimum spanning tree and find its total length.

2 The gardens of a stately home are to be opened to the public. The distances, in metres, between various features are as shown in the table.

	A	B	C	D	E	F	G
A	–	250	200	–	500	300	–
B	250	–	400	200	–	70	
C	200	400	–	300	400	–	300
D	–	200	300	–	–	–	350
E	500	–	400	–	–	–	500
F	300	70	–	–	–	–	–
G	–	–	300	350	500	–	–

(a) Use Prim's algorithm to determine the shortest possible length of pathway to enable visitors to walk between all of the features.

(b) The owners decide to build an ornamental lake between features A and C, so that there is no route between them. What effect does this have on the shortest possible length of pathway?

3 Draw the network which is tabulated in Question 2. Hence apply Kruskal's algorithm, showing your working clearly.

4 A minimum connector algorithm is of quadratic order. Given that a computer implementation of the algorithm takes two seconds to solve a problem with ten nodes, find the approximate length of time that the computer will take to solve a problem with

(a) 20 nodes,   (b) 100 nodes,   (c) 1000 nodes.

5 Adapt Kruskal's algorithm so as to produce an algorithm which will find the spanning tree of *greatest* weight. Apply your algorithm to the network of distances, in kilometres, between towns shown in the diagram.

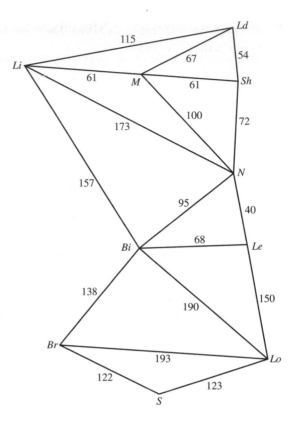

6 The diagram shows a set of holes drilled through a printed circuit board which has a 1 cm grid drawn on it. These need to be joined together using the shortest possible length of conductor. The conductor can only be laid parallel to the sides of the printed circuit board.

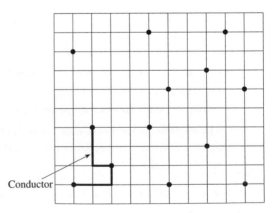

(a) Find the least possible length of conductor, given that two lengths of conductor may only meet at the drilled holes; that is, no new nodes can be created. (Part of the conductor is shown drawn in.)

(b) By how much can the length of conductor be reduced if new nodes may be created?

7 A company has offices in six towns. The costs, in £, of travelling between these towns are shown in the table below.

	A	B	C	D	E	F
A	–	15	26	13	14	25
B	15	–	16	16	25	13
C	26	16	–	38	16	15
D	13	16	38	–	15	19
E	14	25	16	15	–	14
F	25	13	15	19	14	–

(a) Use Prim's algorithm, starting by deleting row A, to find the cheapest way of visiting the six towns. You should show all your working and indicate the order in which the towns were included.

The travel times, in minutes, between the six towns are given in the table below.

	A	B	C	D	E	F
A	–	20	30	20	10	30
B	20	–	20	30	20	10
C	30	20	–	30	30	30
D	20	30	30	–	10	30
E	10	20	30	10	–	10
F	30	10	30	30	10	–

(b) The company wants it to be possible to travel from any town to any other town in under an hour. Show that this is not possible if they just use the arcs from the solution to part (a).

(c) Construct the minimum connector tree for the travel times. Work out how much it would cost to travel from A to C, using just these arcs, and how long it would take.

(OCR)

8 (a) Write down the matrix that represents the network shown in the figure.

(b) Apply Prim's algorithm to this matrix, starting by crossing out row A, to find the arcs that make up a minimum connector for this network. Show all your working clearly, and indicate the order in which you build your minimum connector.

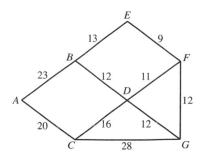

(c) Write down the length of your minimum connector.

(OCR)

4 Finding the shortest path

This chapter is about finding the path between two nodes which has the least weight. When you have completed it you should

- be able to apply the method known as Dijkstra's algorithm
- be aware of the circumstances under which Dijkstra's algorithm can be applied.

4.1 'Shortest'

It is now relatively inexpensive to buy a route planner for a private car. With this you can find the route to your destination which has the shortest distance, or which takes the least time based upon the latest traffic information and expected speeds on different roads. The same algorithm will solve both of these problems; in one case you must consider the network of distances, and in the other case you must consider the network of times. This type of problem is generally known as a 'shortest path' problem but it is better thought of as a problem of finding the path of minimum weight. The weights of arcs may be distances or times, as above, or they could be something entirely different, such as depreciation costs or petrol charges.

The map in Fig. 4.1 shows the main routes between six towns. Single lines are A-roads, and double lines are motorways.

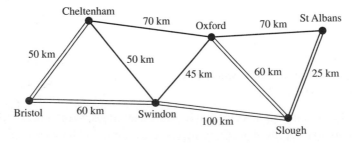

Fig. 4.1

Try to solve the following problems by inspection before looking at the solutions (page 124).

1 What is the shortest path from St Albans to Bristol?

2 Suppose the average speed on the A-roads to be 80 km h^{-1} and on the motorways to be 110 km h^{-1}. What is the quickest route from St Albans to Cheltenham?

3 Suppose further that there is a quarter of an hour delay on the M25 from St Albans to Slough. What now is the quickest route from St Albans to Bristol?

As shortest path questions become more complicated it soon becomes necessary to develop a systematic approach to solving them. One of the most commonly used methods for the shortest path problem is an algorithm invented by Edsger Dijkstra.

4.2 Dijkstra's algorithm

This algorithm is based upon the idea of labelling each node with the length of the shortest path from the start node found so far. This temporary label is replaced whenever a shorter path is found. When you can be certain that there is no shorter route you 'box' the label to show that it is now a **permanent label**.

Dijkstra's algorithm

Step 1   Label the start node with zero and box this label.

Step 2   Consider the node with the most recently boxed label. Suppose this node to be X and let D be its permanent label. Then, in turn, consider each node directly joined to X but not yet permanently boxed. For each such node, Y say, temporarily label it with the lesser of $D+$ (the weight of arc XY) and its existing label (if any).

Step 3   Choose the least of all temporary labels on the network. Make this label permanent by boxing it.

Step 4   Repeat Steps 2 and 3 until the destination node has a permanent label.

Step 5   Go backwards through the network, retracing the path of shortest length from the destination node to the start node.

Example 4.2.1

Find the shortest path from A to G in Fig. 4.2.

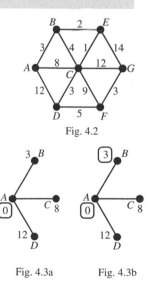

Fig. 4.3a shows the situation after Step 1, in which A was labelled 0 and boxed, and the first pass through Step 2, where B, C and D have temporary labels.

In Step 3, the least of the temporary labels is at B, so this is made permanent by boxing it, as shown in Fig. 4.3b.

Fig. 4.3c shows the situation after the next pass through Step 2, where the temporary label 5 on E comes from adding the most recently boxed label, 3 at B, and the length 2 from B to E. Similarly, the temporary label at C comes from adding the permanent label 3 at B and the length 4 from B to C to make a total of 7. As 7 is less than the existing temporary label 8 at C, the 8 is crossed out and replaced with 7.

Fig. 4.3d shows the label at E boxed, as it is the least of the temporary labels. This provides the starting point for the next pass through Step 2.

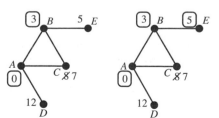

Continuing in this way, you eventually get to the point at which all the labels are permanently boxed. This is shown in Fig. 4.4.

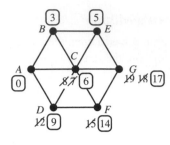

Fig. 4.4

You are now ready for Step 5.

Retracing the path backwards through the network, you can see that

17_G was produced by $14_F + 3$.

14_F was produced by $9_D + 5$, and so on.

The shortest path is $ABECDFG$, which has length 17.

Note that a bonus of Dijkstra's algorithm is that once the algorithm has been carried out you know the shortest paths to all permanently labelled nodes.

If you are required to show the order of permanent labelling, it can be useful to put your working at each node in a box, as shown in Fig. 4.5.

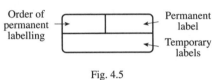

Fig. 4.5

Thus, the box in Fig. 4.6 shows that the relevant node had received an initial temporary label of 8, then one of 7, and finally one of 5. This became its permanent label, and it was the 6th node to be permanently labelled.

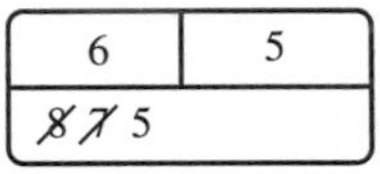

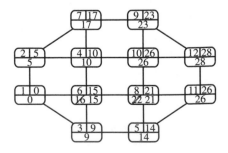

Fig. 4.6

Example 4.2.2
The network shown in Fig. 4.7 represents part of a road system in a city. Some of the roads are one-way, and the weights on the arcs represent estimated travel times, in minutes. What is the quickest route from A to K?

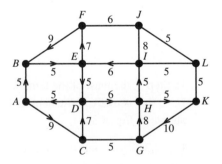

Fig. 4.7

Fig. 4.8 shows the result of applying Dijkstra's algorithm.

Retracing the steps you find that the quickest route is $ABEDHK$, taking a total time of 26 minutes.

Notice that you need not draw a new diagram for each pass through Dijkstra's algorithm.

Fig. 4.8

Exercise 4

1   Use Dijkstra's algorithm to find the shortest path from A to G for the network shown below. Show how you arrive at your result.

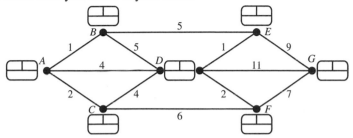

2   Use Dijkstra's algorithm to find the shortest path from A to J for the network below.

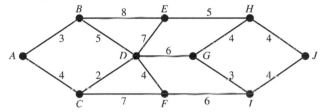

3   The table shows the cost in £ of direct journeys between six towns. Find the minimum cost of travelling from A to F.

	A	B	C	D	E	F
A	–	11	12	–	–	–
B	11	–	–	15	18	–
C	12	–	–	8	21	–
D	–	15	8	–	–	20
E	–	18	21	–	–	15
F	–	–	–	20	15	–

4   Each move of a counter is one square horizontally or vertically on the board shown here. The counter cannot move across the thick lines.

Apply Dijkstra's algorithm to find the smallest number of moves from the square marked with a counter to each other square.

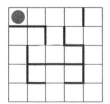

5 Moore's algorithm for a connected graph with two specified nodes, m and n, is as follows.

Step 1   Let $i = 0$.

Step 2   Label m with 0.

Step 3   Find all unlabelled nodes that are adjacent to a node labelled i, and label them $i+1$.

Step 4   Replace i by $i+1$.

Step 5   Repeat Steps 3 and 4 until node n is labelled.

(a)   Apply Moore's algorithm to the graph in the figure.

(b)   What is Moore's algorithm designed to accomplish?

(c)   What is the connection between Moore's algorithm and Dijkstra's algorithm?

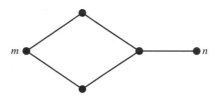

4.3   A drawback of Dijkstra's algorithm

One drawback of Dijkstra's algorithm is that it cannot be used if any weights are negative. (The cost of a route might be negative if, for example, a firm could achieve a profit by making a delivery along that route.)

Example 4.3.1

(a)   What is the shortest route from A to C in the network of Fig. 4.9?

(b)   What is the result of applying Dijkstra's algorithm?

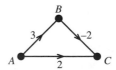

Fig. 4.9

(a)   You can see by inspection that the shortest route is ABC, which has weight 1.

(b)   Fig. 4.10 shows the effect of applying Dijkstra's algorithm. The problem is that node C is permanently labelled before the route ABC is considered.

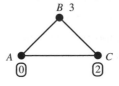

Fig. 4.10

Similarly, Dijkstra's algorithm cannot be used for longest path problems.

4.4* The order of Dijkstra's algorithm

You can, if you wish, omit this section and proceed straight to Miscellaneous exercise 4.

Consider applying Dijkstra's algorithm to a graph with n nodes. At a stage when there are k nodes not yet permanently labelled,

> Step 2 requires carrying out at most k additions and finding the lesser of each of at most k pairs of numbers, which means making k comparisons.

> Step 3 requires finding the least of at most k temporary labels, which means making $k-1$ comparisons.

The total number of additions is therefore at most

$$(n-1)+(n-2)+\ldots+2+1 = \tfrac{1}{2}n(n-1).$$

The total number of comparisons is at most

$$[(n-1)+(n-2)+\ldots+2+1]_{\text{Step 2}} + [(n-2)+(n-3)+\ldots+2+1]_{\text{Step 3}} = (n-1)^2.$$

Dijkstra's algorithm is therefore of order n^2. Thus Dijkstra's algorithm is an efficient algorithm which can be used for relatively large problems without requiring unacceptably large amounts of computing time.

Miscellaneous exercise 4

1 (a) Use Dijkstra's algorithm to find the shortest path from A to F. Show all necessary working.

   (b) A directed arc from C to D is now added, with weight -4. What is the shortest path from A to F now? Explain why Dijkstra's algorithm cannot be used to find this path.

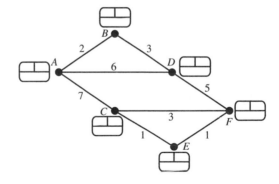

2 The numbers on the diagram represent the times in minutes of the journeys between railway stations.

   (a) What is the quickest route from A to D?

   (b) How could you adjust the numbers to model the fact that there is a delay of 20 minutes on all journeys passing through station O? What is now the quickest route from A to D?

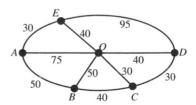

3 It is required to find the shortest paths from each of A, B and C to N.

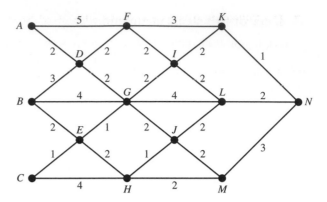

(a) How can these paths be determined by applying Dijkstra's algorithm just once?

(b) Obtain the three shortest paths. Which of A, B and C is nearest to N?

4 Fares (in £) for direct flights between five cities are shown in the table.

	A	B	C	D	E
A	–	90	70	35	30
B	90	–	40	150	55
C	70	40	–	20	50
D	35	150	20	–	100
E	30	55	50	100	–

(a) Draw a network and use Dijkstra's algorithm to find the cheapest routes from A to each other city.

(b) Suppose each change of flight is estimated to cost an extra £10 of sundry expenses. Find the cheapest routes now.

5 The numbers on the arcs of this network represent the maximum weight (in tons) of a vehicle allowed on the road that the arc represents.

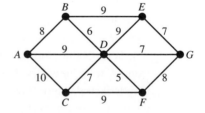

(a) What is the heaviest vehicle that can travel from A to G?

(b) How can Dijkstra's algorithm be modified to solve problems of this type?

6 Apply Dijkstra's algorithm to find the shortest path from A to I.

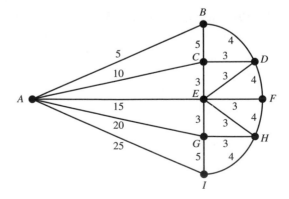

7 Use Dijkstra's algorithm to find the length of the shortest path from A to B.

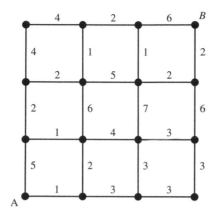

8 Suppose you purchase a new computer which is 100 times as fast as your old one. What gain in problem size per hour would you achieve in the use of Dijkstra's algorithm? Show all necessary working.

9 The owner of a stately home is concerned that wheelchair access to his property is limited. Currently it is impossible to visit the upper floor, the servants' quarters, the chapel, the lake and the tower in a wheelchair; all other parts of the house and its grounds are accessible.

The costs (in £1000's) of installing ramps or lifts to connect the various parts of the property are shown in the table below. X means that a direct connection is not possible.

	House and grounds	Upper floor	Servants' quarters	Chapel	Lake	Tower
House and grounds	–	3	5	7	7	6
Upper floor	3	–	1	X	X	2
Servants' quarters	5	1	–	2	4	X
Chapel	7	X	2	–	6	3
Lake	7	X	4	6	–	1
Tower	6	2	X	3	1	–

(a) Draw a network to represent this information.

(b) Use Dijkstra's algorithm to find the cheapest way of connecting each of the five areas to the house and grounds. Show all your working clearly, and indicate the order in which you assign permanent labels to nodes.

(c) Use the solution to part (b) to list which connections should be made for wheelchair users to be able to reach all parts of the property, at minimum cost to the owner. Explain how you knew which arcs to include, and find the cost of making these connections.

(d) Explain what is special about the arcs that make up the solution, and name an algorithm that could have been used to find the solution in a more direct way.   (OCR)

10 Little Red Riding Hood wants to travel through the woods from her house, at point A, to Grandma's house, at point G. The network below shows the possible paths that she may use, and the time (in minutes) to travel each path.

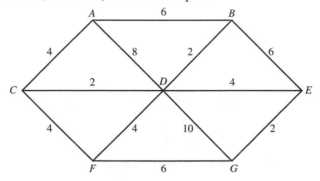

(a) Use Dijkstra's algorithm to find the quickest route from A to G. Show all your working clearly, and indicate the order in which you assign permanent labels to nodes.

(b) The big bad wolf also wants to travel from A to G. He knows which route Little Red Riding Hood has chosen, and he must avoid using any of the arcs on her route.

Find the quickest route from A to G for the big bad wolf.

(c) The big bad wolf travels twice as quickly as Little Red Riding Hood. Assuming that Little Red Riding Hood and the big bad wolf leave A at the same time, and that they use the routes found above, work out how long the big bad wolf will have to wait at Grandma's house before Little Red Riding Hood arrives.   (OCR)

11 The table shows the distances (in metres) of the shortest direct route between each of seven classrooms, A, B, C, D, E, F and G.

	A	B	C	D	E	F	G
A	0	200	50	35	65	80	250
B	200	0	150	200	120	100	25
C	50	150	0	10	5	25	150
D	35	200	10	0	20	40	200
E	65	120	5	20	0	10	100
F	80	100	25	40	10	0	100
G	250	25	150	200	100	100	0

(a) Use Dijkstra's algorithm to find the shortest distance from classroom A to classroom B. Show all your working clearly, and indicate the order in which you assign permanent labels to the nodes.

The teacher in classroom A asks a pupil to take a message to each of the other classrooms, and then come back to classroom A.

(b) Use your answer to part (a) to find an upper bound for the shortest route that the pupil can take.   (OCR)

12 The figure shows a network of the main routes between Penzance and Winchester. The values on the arcs represent the distances in miles. The thinner arcs represent 'A' roads, and the thick arcs represent motorways.

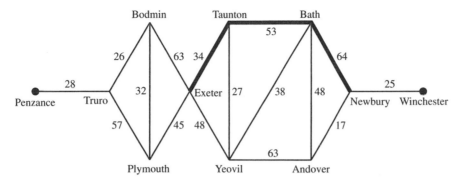

(a) Use Dijkstra's algorithm for the whole network to find

   (i) the shortest route from Exeter to Penzance, and the length of this route,

   (ii) the shortest route from Exeter to Winchester, and the length of this route.

Show all your working clearly, and indicate the order in which you assign permanent labels to the towns.

(b) A motorist can average 48 miles per hour on 'A' roads and 66 miles per hour on motorways. Calculate the shortest time (in hours and minutes) that it takes the motorist to drive from Penzance to Winchester. (OCR)

13 A couple are eloping to Gretna Green when they notice that the cars ahead of them have come to a complete standstill because of an accident. They turn off the main road and consult the map.

The network in the figure represents the roads that the couple can use to get from the accident (A) to Gretna Green (G). The length of each section of road is shown in miles.

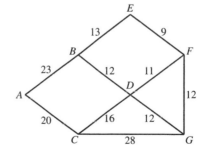

(a) Use Dijkstra's algorithm, on a copy of the figure, to find the shortest route from A to G. Show all your working clearly, and indicate the order in which you assign permanent labels to the nodes.

Before reaching B, the couple decide instead to stop at F for the night.

(b) Write down the shortest route from A to F. (OCR)

5 Route inspection

This chapter looks at the problem of finding a closed trail covering every arc of a network. When you have completed it you should

- be aware of a wide range of route inspection problems
- be able to apply a standard method of solution to obtain the closed trail of minimum weight.

5.1 Traversability

Consider the task of a village police officer who must traverse each of the streets shown in Fig. 5.1.

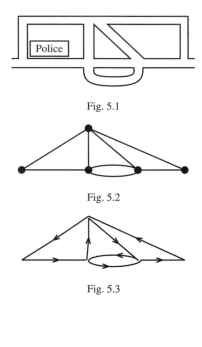

Fig. 5.1

Replacing roads by arcs and junctions by nodes, you obtain the graph shown in Fig. 5.2.

Since each node has even order, the graph is Eulerian. The police officer can therefore choose a closed path which contains every arc precisely once, as shown in Fig. 5.3.

Fig. 5.2

The length of the police officer's route is then precisely the same as the length of all the streets, and that is clearly the best that can be achieved. However, for most networks of roads the situation will not be so simple.

Fig. 5.3

For example, consider Fig. 5.4.

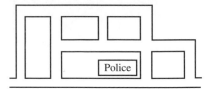

Fig. 5.4

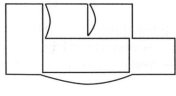

Fig. 5.5

Here, the underlying graph has six nodes of odd order. The only way that a closed tour can traverse *every* arc is if some arcs are repeated. For example, Fig. 5.5 shows that three arcs are traversed twice.

The length of the police officer's route can now be thought of as

the length of all the streets + the lengths of repeated streets.

There are many practical situations which require the solution of a route inspection problem (that is, the finding of a closed tour containing every arc at least once). Situations similar to that of the police patrol are repairing track, delivering mail, seeding fields and clearing snow.

A modern application of route inspection is the checking of every link on a web site. Web sites can have hundreds of pages and thousands of links, and checking these links requires the involvement of both computers and humans. Computer software is used to check simple links and to keep track of the whole checking process, whereas human input is especially required for checking links which are more descriptive.

5.2 The Chinese Postman algorithm

You have seen that modern route inspection problems can involve large networks. There is therefore a need for a systematic procedure to obtain a closed trail containing every arc of minimum length, or weight.

The following well known procedure for finding the least-weight closed trail containing every arc (as required by a postman) was invented by a Chinese mathematician, Kuan Mei-Ko, in 1962.

Chinese Postman algorithm

Step 1   Find all nodes of odd order.

Step 2   For each pair of odd nodes find the connecting path of minimum weight.

Step 3   Pair up all the odd nodes so that the sum of the weights of the connecting paths from Step 2 is minimised.

Step 4   In the original graph, duplicate the minimum weight paths found in Step 3.

Step 5   Find a trail containing every arc for the new (Eulerian) graph.

Example 5.2.1

Find the minimum-weight closed trail containing all arcs for the network in Fig. 5.6.

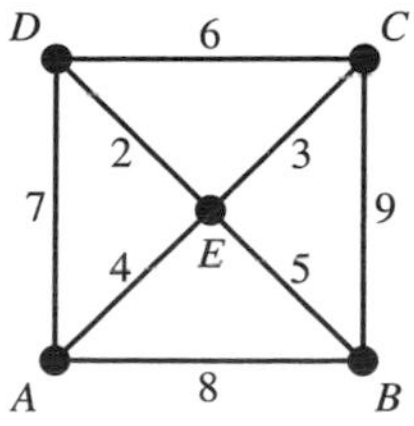

Fig. 5.6

The odd nodes are A, B, C and D (Step 1).

The minimum weights of the connecting paths are

$$AB\ 8,\quad BC\ 8,\quad AC\ 7,\quad BD\ 7,\quad AD\ 6,\quad CD\ 5\ \text{(Step 2)}.$$

The possible pairs, in which all the odd nodes are connected, are

$$AB,\ CD\ 8+5=13,\quad AC,\ BD\ 7+7=14,\quad AD,\ BC\ 6+8=14.$$

So the odd nodes should be paired up as AB and CD (Step 3).

Add in the arcs *AB* and *CED* (Step 4).

Step 5 gives a trail of minimum weight: for example, *ABAEBCECDEDA*.

Its weight is

the length of all the streets + the lengths of repeated streets = 44 + 13 = 57.

The Chinese Postman algorithm assumes that the original graph has an even number of nodes of odd order. This is actually true for *all* graphs. Note that each arc of a graph contributes 1 to the orders of two nodes, and so the sum of the orders of all the nodes is twice the number of arcs. The number of nodes of odd order must therefore be even, as required. (See also Exercise 2C Question 7.)

When the number of nodes of odd order is relatively large, it can be very time-consuming to check all the possibilities, as in the next example.

Example 5.2.2
The network given in Fig. 5.7 gives the times (in minutes) it takes to drive along a number of streets in central London.

(a) List the nodes of odd order.

(b) Draw up a table showing the least times between the nodes of odd order.

(c) What would be the minimum time needed to complete a closed trail of every one of these streets? Write down such a closed trail.

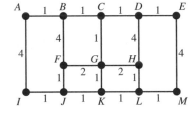

Fig. 5.7

(a) *B, C, D, F, H, J, K, L.*

(b)

	B	C	D	F	H	J	K	L
B	–	1	2	4	4	4	3	4
C	1	–	1	3	3	3	2	3
D	2	1	–	4	4	4	3	4
F	4	3	4	–	4	1	2	3
H	4	3	4	4	–	3	2	1
J	4	3	4	1	3	–	1	2
K	3	2	3	2	2	1	–	1
L	4	3	4	3	1	2	1	–

(c) The total weight of all streets is 32.

For Step 3, one good possibility for pairing off the nodes is

BD, CK, FJ, HL.

These paths have total length $2+2+1+1=6$.

For this to be beaten or even equalled, no arc of length 4 can be used, since $1+1+1+4>6$. Thus the extra path from B *must* start BC and the extra path from D *must* start DC. It should now be clear that the extra 6 minutes cannot be beaten and so the minimum time is $32+6=38$ minutes. An example of a minimum time trail is

AIJFJ KLHLM EDHGF BCDCG KGCBA.

Exercise 5

1 A person delivering leaflets has to walk along each of the roads shown on the map. All the measurements are metres, and, except for at the crescent, all the angles are right angles.

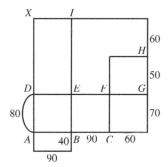

  (a) Explain the relevance of those intersections where an odd number of roads meet.

  (b) Find the shortest possible distance the person has to walk, starting and finishing at X. Show the results of each possible pairing of odd nodes.

  (c) For each of the nine road intersections, find the number of times that the delivery person will pass through that intersection.

  (d) What is the shortest possible distance the delivery person would have to walk if they started and finished at C? Explain your answer.

2 (In this question ignore the widths of the roads.)

  (a) Model this road system as a network in an appropriate way for

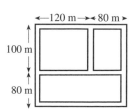

    (i) a delivery man who only needs to walk along each road once,

    (ii) a postwoman who walks along both sides of each road,

    (iii) a street cleaner who travels along both sides of each road in the correct direction.

  (b) What is the total length of the roads in the network?

  (c) In each case considered in part (a), find the total distance that will need to be travelled.

3 An electronic game takes place in an arena consisting of a 4×4 block of connected rooms.

To complete the game, it is necessary to travel through each of the 24 doorways. Convert this problem into a route inspection problem, and find the least number of doorways that will have to be gone through twice.

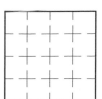

4[*] (a) Consider the problem in Question 3, but with an $n \times n$ block of rooms, where n is even. The rooms are connected as in Question 3. Copy and complete the table.

Order		2	3	4
Number of nodes with that order				

How many arcs (doorways) will need to be repeated?

(b) How many arcs must be repeated if n is odd?

5.3 Pairing odd nodes

You have seen that the version of the Chinese Postman algorithm given in this chapter depends upon your performing the following operations.

- Consider all possible pairings of the nodes of odd order.
- Find the shortest distance between each pair of nodes of odd order.

The second of these can be performed by Dijkstra's algorithm, which has quadratic order. The important question, therefore, is how many times must this algorithm be performed, that is, how many pairings are there of the nodes of odd order.

For two odd nodes A and B, the only arc is AB, so there is 1 pairing.

For four odd nodes, A, B, C and D, the possible choices are AB with CD, AC with BD and AD with BC, so there are 3 pairings.

For six odd nodes, A, B, C, D, E and F,

you can choose AB along with each of the three pairings of C, D, E and F;
you can choose AC along with each of the three pairings of B, D, E and F;
you can choose AD along with each of the three pairings of B, C, E and F;
you can choose AE along with each of the three pairings of B, C, D and F;
you can choose AF along with each of the three pairings of B, C, D and E.

There are therefore 15 pairings in total.

Table 5.8 shows the results for various numbers of nodes of odd order.

Number of odd nodes	Number of ways of pairing
2	1
4	3
6	15
8	105
10	945
12	10 395
14	135 135

Table 5.8

For more than six nodes of odd order it is clearly impractical to apply the Chinese Postman algorithm by hand. Even using a computer is problematical if the number of odd nodes is large, because the number of pairings is so large.

Exercise 5B

1   A child's toy consists of a number of pegs around which elastic bands are wrapped.

(a)    (b)

Which of (a) and (b) can be made with a single, continuous elastic band? Explain your answer.

2   A snow-plough must drive along all the main roads shown, starting and finishing at the garage at A. The distances in kilometres are marked.

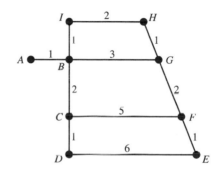

(a)   Explain why 25 km is a lower bound for the distance the snow-plough must travel.

(b)   Find the least distance it must actually travel, showing your method clearly.

3   The matrix form for a weighted graph is shown in the table below.

	A	B	C	D	E	F	G	H
A	–	23	17	–	18	–	15	–
B	23	–	9	10	12	16	–	14
C	17	9	–	9	20	–	27	–
D	–	10	9	–	–	–	–	16
E	18	12	20	–	–	7	20	–
F	–	16	–	–	7	–	24	17
G	15	–	27	–	20	24	–	–
H	–	14	–	16	–	17	–	–

(a)   Use Dijkstra's algorithm to find the shortest paths from H to C, D and E.

(b)   Draw a weighted graph with nodes C, D, E, H and with each arc having the weight of the corresponding shortest path.

(c)   Apply the Chinese Postman algorithm to the original network. Which arcs should be duplicated?

4   (a)   Look at Table 5.8. Why is the number of ways of pairing 8 nodes of odd order 105?

(b)   Find a general formula for the other results in Table 5.8.

5 The following table of distances (in metres) between features in the gardens of a stately home was considered in Miscellaneous exercise 3.

	A	B	C	D	E	F	G
A	–	250	200	–	500	300	–
B	250	–	400	200	–	70	–
C	200	400	–	300	400	–	300
D	–	200	300	–	–	–	350
E	500	–	400	–	–	–	500
F	300	70	–	–	–	–	–
G	–	–	300	350	500	–	–

(a) What is the total length of all the paths?

(b) What distance must be covered if each path is to be inspected?

6 Apply the Chinese Postman algorithm to the network of distances, in kilometres, shown below. Which roads should be repeated?

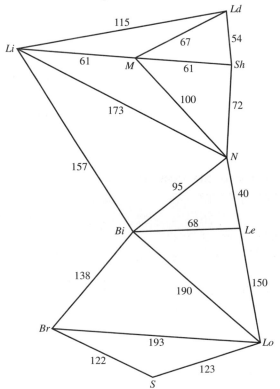

7 A network of pipelines is as shown, with distances in metres. A fault has to be located. What is the shortest route which will cover every length of pipeline at least once? Show your method fully.

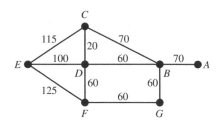

Miscellaneous exercise 5

1 A steam railway connects four stations, as shown in the figure.

Ann O'Rak, a steam railway enthusiast, wants to travel every section of track, starting and finishing at A.

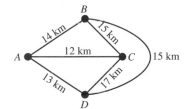

(a) Explain why it will be necessary for Ann to travel some sections of track twice. Write down the minimum number of tracks that will have to be repeated.

(b) By considering all pairings, find the minimum distance that Ann must travel to cover every section of track, starting and finishing at A. Give a possible route that she could take.

(c) Suppose that the railway had connected five stations, with every station connected to every other station. How many pairings would need to be considered to solve Ann's problem? (OCR)

2 (a) Explain why the number of odd nodes in any graph is always an even number.

The network in the figure represents the paths in a woodland park; the distances are in hundreds of metres.

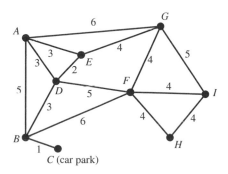

The members of a family have been walking in the woods. When they get back to the car park they find that their youngest child has lost his teddy bear somewhere on the walk. They do not know which paths they have used, so the father decides to go back and cover every path in the woods.

(b) By considering all possible pairings of odd nodes, find the length of the shortest route that the father can take to cover every path in the woods. You should explain your method carefully. (OCR)

3 The figure shows the bipartite graph $K_{3,3}$, in which each of the three nodes A, B and C are joined to each of the three nodes X, Y and Z using the nine arcs shown.

The table gives the length, in kilometres, of each arc.

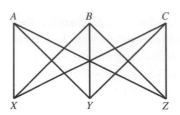

	X	Y	Z
A	2	3	4
B	15	11	12
C	6	5	9

Consider the route inspection problem of finding the length of the shortest route that starts and finishes at A, and travels each arc at least once. The method to be used is to check all the possible pairings between the nodes on one side of the bipartite graph and the nodes on the other side of the bipartite graph.

(a) Apply the method to the graph shown in the figure.

(b) Show that using the method on the graph $K_{3,3}$ requires $(2 \times 6) + 8 + 1 = 21$ additions.

In the bipartite graph $K_{5,5}$, each of the five nodes A, B, C, D and E is joined to each of the five nodes, V, W, X, Y and Z using twenty-five arcs.

(c) Show that using the method on the graph $K_{5,5}$ requires $(4 \times 120) + 24 + 1 = 505$ additions.

(d) Calculate the number of additions needed when using the method on $K_{n,n}$ where n is an odd positive integer.

(e) Use your answer to part (d) to explain how you know that the method cannot be a polynomial order algorithm. (OCR, adapted)

4 A highways department has to inspect its roads for fallen trees.

(a) The diagram shows the lengths of the roads, in miles, that have to be inspected in one district.

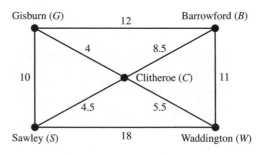

    (i) List the three different ways in which the four odd vertices in the diagram can be paired.

    (ii) Find the shortest distance that has to be travelled in inspecting all the roads in the district, starting and finishing at the same point.

(b) The connected graph of the roads in another district has six odd vertices. Find the number of ways of pairing these odd vertices.

(c) For a connected graph with n odd vertices, find an expression for the number of ways of pairing these odd vertices. (AQA)

5 The edges of the network in the figure represent roads in a small housing estate. There is only one road into and out of the estate, represented by edge AB. The lengths of the roads are shown in metres.

The total length of roads is 2300 m.

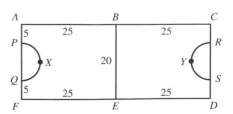

(a) A papergirl delivers to about 20% of the houses and therefore finds it worthwhile to cross from side to side of a road whilst making deliveries. Thus, she would prefer to walk along each road only once.

Use an appropriate algorithm to find the minimum total length of road along which she must walk if she starts and finishes her round at A. Indicate how you applied the algorithm.

(b) A postman delivers to about 80% of the houses. He finds it better to deliver to one side of a road at a time. He therefore needs to walk along each road at least twice. If he starts and finishes at A, find the minimum distance that he must walk, justifying your answer.

(c) Suppose that the postman acquires a bicycle. This means that he would still like to travel along each road twice, but in opposite directions, so that he is always riding on the correct side of the road. Will this mean that he must travel further? Justify your answer.   (AQA)

6 A groundsman at a local sports centre has to mark out the lines of several five-a-side pitches using white paint. He is unsure as to the size of the goal area and he decides to paint the outline as given below, where all the distances given are in metres.

(a) He starts and finishes at the point A. Find the minimum total distance that he must walk and give one of the corresponding possible routes.

(b) Before he starts to paint the second pitch he is told that each goal area is a semicircle of radius 5 metres, as shown in the diagram below.

   (i) Find an optimal 'Chinese postman' route around the lines. Calculate the length of your route.

   (ii) State which vertices would be suitable starting points to keep to a minimum the total distance walked from when he starts to paint the lines until he completes the painting.

walked from when he starts to paint the lines until he completes the painting.

(AQA)

7 A road gritting service is based at a point A. It is responsible for gritting the network of roads shown in the diagram, where the distances shown are in miles.

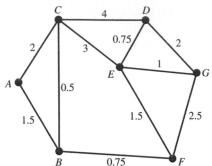

(a) Explain why it is not possible to start from A and, by travelling along each road only once, return to A.

(b) In the network there are four odd vertices, B, D, F and G. List the different ways in which these odd vertices can be arranged as two pairs.

(c) For each pairing you have listed in part (b), write down the sum of the shortest distance between the first pair and the shortest distance between the second pair.

(d) Hence find an optimal 'Chinese Postman' route around the network, starting and finishing at A. State the length of your route.   (AQA)

8 (a) The network represents a road system in which the lengths of the roads are shown in kilometres. The road system has to be cleared of snow by a snow-plough which is based at A. The snow-plough only needs to travel along each road once in order to clear it. However, when all roads have been cleared it must return to its base at A.

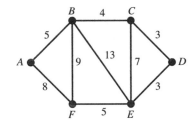

(i) State which roads the snow-plough should drive along twice in order to travel the minimum total distance.

(ii) Hence solve the Chinese Postman problem for this network to find a route for the snow-plough, starting and finishing at A.

(iii) The road BE becomes a dual carriageway and must be cleared twice, once in each direction. Redraw the network to take account of this and find, by inspection, a new route for the snow-plough.

(b) The original network has to be drawn as efficiently as possible by a pen operated by a computer. Explain how this can be done without lifting the pen from the paper and without tracing any of the edges more than once.   (AQA)

6 The travelling salesperson problem

This chapter is about finding a tour that visits every node of a network. When you have completed it you should

- appreciate that evaluating all tours is not practical for large scale problems
- be able to apply an algorithm to find a solution (not necessarily the best)
- know how to obtain bounds within which the best solution must lie.

6.1 The classical problem

You have seen that the problem of finding a tour that visits every arc (the route inspection problem) depended upon the earlier work of the mathematician Euler. Similarly, the problem of finding a tour visiting every node was studied in the 19th century by the Irish mathematician Sir William Hamilton.

A **Hamiltonian cycle** is defined to be a tour which contains every node precisely once. In a simple case, such as that of the network in Fig. 6.1, it is easy to list all the Hamiltonian cycles.

There are just three essentially different Hamiltonian cycles:

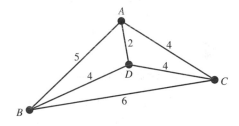

Fig. 6.1

> $ACBDA$ with weight 16,
> $ABCDA$ with weight 17,
> $ABDCA$ with weight 17.

Note, for example, that the cycle ADBCA is just the first cycle reversed and so is not essentially different from it.

The classical travelling salesperson problem is to find the Hamiltonian cycle of minimum weight. In the above case this is the cycle $ACBDA$.

However, not all graphs have Hamiltonian cycles. For example, the network shown in Fig. 6.2 does not have any cycles passing through A.

Nevertheless, a salesperson living, say, in town B might still need to find the shortest round trip visiting every town. To enable you to use this chapter's methods for finding Hamiltonian cycles, you can replace any network like Fig. 6.2 by the complete network of shortest distances. The shortest distance between A and C is 33, via D. If you add the direct arc AC, of weight 33, you have not changed the

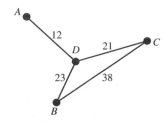

Fig. 6.2

problem. Adding all such arcs, you get Fig. 6.3.

For the remainder of this chapter it is therefore assumed that the problem is always the classical one of finding a Hamiltonian cycle of minimum weight, with no repetition of nodes.

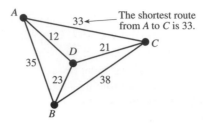

Fig. 6.3

6.2 A difficult problem

In Section 6.1 you saw how to find the minimum Hamiltonian cycle for a graph with four nodes by listing all three possible cycles.

Unfortunately, as the number of nodes increases, the number of possible Hamiltonian cycles tends to increase very rapidly. For graphs where all nodes are directly linked to each other, Table 6.4 shows the numbers of Hamiltonian cycles for small values of n.

Number of nodes, n	Number of Hamiltonian cycles
3	1
4	3
5	12
6	60
7	360

Table 6.4

To see how to calculate these numbers, consider the case $n = 5$ and suppose the nodes are A, B, C, D and E. Node A must be on each cycle, and so you may as well always start from A. There are then four possibilities for the next node, three for the one after, two for the one after that, and just one possibility for the final node before returning to A.

This gives $4 \times 3 \times 2 \times 1 = 24$ sequences. However, each Hamiltonian cycle corresponds to two sequences, since, for example, $ABCDEA$ and its reverse $AEDCBA$ are taken to be the same Hamiltonian cycle. So if $n = 5$, there are $\frac{1}{2} \times 24$, or 12, Hamiltonian cycles.

In general, for n nodes, the number of Hamiltonian cycles is

$$\tfrac{1}{2} \times (n-1) \times (n-2) \times \ldots \times 3 \times 2 \times 1 = \tfrac{1}{2}(n-1)!$$

The symbol $n!$, called 'factorial n', denotes $n \times (n-1) \times (n-2) \times \ldots \times 3 \times 2 \times 1$. You will find it in P2 Section 3.4.

The method of evaluating all Hamiltonian cycles therefore requires consideration of $\frac{1}{2}(n-1)!$ cycles, and so has exponential order. To see what this means for computer time, consider a 600 MHz processor (that is, one capable of performing 600 million simple operations per second). If there were 20 nodes, there would be $\frac{1}{2} \times 19!$, or 6×10^{16}, Hamiltonian cycles. Even with the assumption that the computer could check an entire tour in a single operation, a problem with just 20 nodes would require about 10^8 seconds, which is about 3 years, of computer time!

All methods discovered to date for solving the travelling salesperson problem have exponential order and so attention has been focused on finding not the optimal solution, but simply a reasonably good solution.

6.3 The Nearest Neighbour algorithm

A simple way of trying to find a reasonably good Hamiltonian cycle is to try a greedy algorithm, such as the Nearest Neighbour algorithm. At each stage it visits the nearest node which has not already been visited.

The Nearest Neighbour algorithm

Step 1 Choose any starting node.

Step 2 Consider the arcs which join the previously chosen node to not-yet-chosen nodes. From these arcs pick one that has minimum weight. Choose this arc, and the new node on the end of it, to join the cycle.

Step 3 Repeat Step 2 until all nodes have been chosen.

Step 4 Then add the arc that joins the last-chosen node to the first-chosen node.

Example 6.3.1

A small chemical plant can be used to produce any one of five chemicals, A, B, C, D and E. The times (in hours) required for cleaning the equipment, and setting it up again for making the next chemical, are as shown in Fig. 6.5. Use the Nearest Neighbour algorithm to find a small total changeover time for production of all five chemicals, starting and finishing with chemical A.

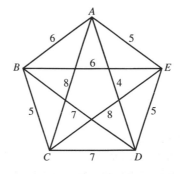

Fig. 6.5

Starting from A, the first arc is AD, as 4 is the least of 6, 8, 4 and 5.

From D, DE is chosen as 5 is the least of 7, 7 and 5.

This is followed by EB, with weight 6 as the lesser of 6 and 8.

Finally, you are forced to choose BC and CA.

The cycle is $ADEBCA$, which has weight 28 hours.

The Nearest Neighbour algorithm is 'greedy' because at each stage the immediately best route is chosen, without a look ahead to possible future problems. The next example shows how this greed can sometimes lead to a poor solution.

Example 6.3.2

A warehouse in Toulouse supplies goods to retail outlets in Bordeaux, Calais, Dijon, Lyons, Marseille, Orléans, Poitiers and St-Etienne. The distances involved (in kilometres) are shown in Fig. 6.6. Use the Nearest Neighbour algorithm, starting from Toulouse, to find a single delivery route to all the towns.

Bordeaux								
870	Calais							
641	543	Dijon						
550	751	192	Lyons					
649	1067	507	316	Marseille				
457	421	297	445	761	Orléans			
247	625	515	431	733	212	Poitiers		
519	803	244	59	309	392	421	St-Etienne	
244	996	726	535	405	582	435	528	Toulouse

Fig. 6.6

The solution is

$$\text{Toulouse} \xrightarrow{244} \text{Bordeaux} \xrightarrow{247} \text{Poitiers} \xrightarrow{212} \text{Orleans} \xrightarrow{297} \text{Dijon}$$

$$\downarrow 192$$

$$\text{Toulouse} \xleftarrow{996} \text{Calais} \xleftarrow{1067} \text{Marseille} \xleftarrow{309} \text{St-Etienne} \xleftarrow{59} \text{Lyons}$$

You should notice that because it was greedy early on, the tour has had to include two extremely large distances at the end. A glance at a map of France shows that Calais should have been included somewhere near the Poitiers–Orléans–Dijon stretch of the tour. Sometimes, a different choice of initial node avoids this type of problem (see Exercise 6A Question 2).

Another possible problem with the Nearest Neighbour algorithm is that it may lead to an incomplete tour, as you will see in the next example.

Example 6.3.3

A holidaymaker on Guernsey hires a bicycle at St Peter Port and wants to complete a tour of the nine places marked on the map of Fig. 6.7. What happens if the Nearest Neighbour algorithm is applied, starting at St Peter Port?

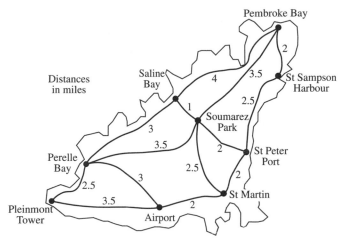

Fig. 6.7

Depending upon the first choice of arc from St Peter Port, the algorithm starts by giving either

St Peter Port–Soumarez Park–Saline Bay–...

or

St Peter Port–St Martin–Airport–Perelle Bay–... .

In both cases, you have crossed from one side of the island to the other. If you now go south, you cannot complete the tour by visiting the places in the north without re-visiting one of the places on the route which crosses the island. Similarly, if you go north, you also have to re-visit one of the places on the route which crosses the island to get to the south of the island.

In conclusion, you should treat the Nearest Neighbour algorithm as only a rough and ready attempt to obtain a good tour. Do not worry about the fact that it does not always lead to a good solution.

Exercise 6

1 (a) List the three possible Hamiltonian cycles for the network of Fig. 6.3.

   (b) What route should the salesperson take for the network shown in Fig. 6.2?

2 Apply the Nearest Neighbour algorithm to the problem of Example 6.3.2, but this time starting from Calais.

3 Delete the road from Perelle Bay to the Airport on the map given in Fig. 6.7. Show how the Nearest Neighbour algorithm, starting at Pembroke Bay, can now lead to a Hamiltonian cycle for this network. What is its length?

4 A delivery firm's costs (in £) for travelling between five towns are as shown in the table.

	A	B	C	D	E
A	–	60	50	40	70
B	60	–	90	–	80
C	50	90	–	80	–
D	40	–	80	–	90
E	70	80	–	90	–

(a) Find the cost for a round trip through all the towns by using the Nearest Neighbour algorithm starting from A.

(b) Find an improved route by using the Nearest Neighbour algorithm starting from a different town.

(c) From which town does the Nearest Neighbour algorithm not work?

6.4 A lower bound

You have seen that even when the Nearest Neighbour algorithm does lead to a Hamiltonian cycle, this may well not have the minimum possible weight. How, therefore, do you know whether a cycle that you have found is close to being the best possible, or whether you should continue searching for a much better one? Fortunately, there is a clever method for showing that there is a limit to how low the total weight of a Hamiltonian cycle can be.

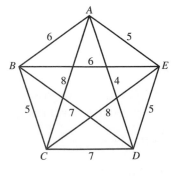

Fig. 6.8

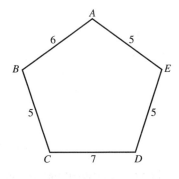

Fig. 6.9

In Example 6.3.1, a Hamiltonian cycle of total weight 28 was found for the network shown in Fig. 6.8. Consider this Hamiltonian cycle as drawn in Fig. 6.9.

You should be able to see that *any* Hamiltonian cycle for the original network will consist of

- two arcs from node A, and
- three arcs linking the points B, C, D, E.

Note that the two arcs from node A must have a total weight at least $4+5$ (taking the two smallest weights of arcs incident at A). Furthermore, by applying Prim's algorithm to the network on just the points B, C, D and E, you can see that the three arcs linking the points B, C, D and E must have total weight at least $5+5+6$ (the weight of a minimum connector on B, C, D and E).

So it is *impossible* for any Hamiltonian cycle to have total weight less than $4+5+5+5+6=25$.

You now know that the minimum possible weight for a Hamiltonian cycle for this network lies between 25 hours and the value found in Example 6.3.1, 28 hours.

The upper bound of 28 hours was obtained by simply finding a tour with that weight. The lower bound of 25 hours was found by considering the network on B, C, D and E separately from the arcs incident at A.

Either by finding a different tour or by splitting the network up differently, or both, it may be possible to further 'improve' these bounds. That is, it may be possible to narrow the range within which the weight of the minimum Hamiltonian cycle must lie.

The method for finding a lower bound can be described in general as an algorithm.

The Lower Bound algorithm

Step 1 Choose an arbitrary node, say X. Find the total of the two smallest weights of arcs incident at X.

Step 2 Consider the network obtained by ignoring X and all arcs incident to X. Find the total weight of the minimum connector for this network.

Step 3 The sum of the two totals is a lower bound.

Example 6.4.1
Consider the network shown in Fig. 6.8.
(a) Apply the Nearest Neighbour algorithm starting from B.
(b) Apply the Lower Bound algorithm with C as the special node.
(c) What can you say about the minimum Hamiltonian cycle?

(a) The algorithm gives $B \xrightarrow{5} C \xrightarrow{7} D \xrightarrow{4} A \xrightarrow{5} E \xrightarrow{6} B$, with weight 27.

(b) For C, the sum of the two smallest arcs is $5+7=12$.

For the remainder, the minimum connector is $4+5+6=15$.

The total is therefore $12+15=27$.

(c) The solutions to (a) and (b) show that $27 \leqslant$ minimal weight $\leqslant 27$. So in this case the Nearest Neighbour algorithm has found the minimum Hamiltonian cycle.

6.5 Tour improvement

In cases where there is a large gap between the upper and lower bounds for a tour, it may be worth trying to improve the best tour obtained so far, rather than looking for a completely different tour. A number of algorithms have been developed to attempt this improvement. In this section you will learn one such method.

Consider, for example, the network shown in Fig. 6.10 with the weight on each arc being the distance on the page between the two points.

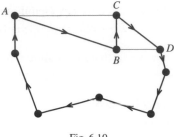

Fig. 6.10

Applying the Nearest Neighbour algorithm from the node B would lead to the tour shown in Fig. 6.10. However, the part of the tour $ABCD$ can be replaced by $ACBD$. This gives a reduction in distance, because

$$d(A,C) + d(B,D) < d(A,B) + d(C,D),$$

where $d(V,W)$ means the weight of the arc between nodes V and W.

You can use this idea as the basis for a general attempt to improve tours. Let V_1, V_2, ..., V_n be the successive nodes of a Hamiltonian tour, and let $V_{n+1} = V_1$, $V_{n+2} = V_2$ and $V_{n+3} = V_3$.

Tour Improvement algorithm

Step 1　Let $i = 1$.

Step 2　If $d(V_i, V_{i+2}) + d(V_{i+1}, V_{i+3}) < d(V_i, V_{i+1}) + d(V_{i+2}, V_{i+3})$, then swap V_{i+1} and V_{i+2}.

Step 3　Replace i by $i+1$.

Step 4　If $i \leqslant n$ then go back to Step 2.

To apply an algorithm such as this one to a large tour may be time-consuming. However, it can easily be programmed for a computer.

When using a computer program to improve tours it can be useful to start with an initial tour, even a very inefficient one. One possibility is to travel along each arc of a minimum connector twice, as in the next example. Note that this will give you a closed trail but not a cycle; a Hamiltonian cycle can then be obtained by finding short-cuts.

Example 6.5.1
Consider the network in Fig. 6.11, which is a copy of Fig. 6.1.

(a)　Find a minimum connector, and hence find a closed trail containing all the nodes with total weight twice that of the minimum connector.

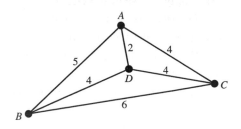

Fig. 6.11

(b)　Obtain a Hamiltonian cycle by finding short-cuts, and then apply the Tour Improvement algorithm to obtain a tour of weight 16.

(a) Fig. 6.12 shows a minimum connector, and Fig. 6.13 shows a closed trail *ADCDBDA* whose weight is twice that of the minimum connector.

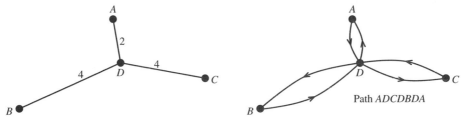

Fig. 6.12   Fig. 6.13

(b) The tour can be reduced by going from *A* to *C* directly instead of via *D*, and by going from *B* to *A* directly instead of via *D*. The tour is then *ACDBA*, of weight 17.

When you do Step 2 of the Tour Improvement algorithm with $i = 2$, you find that

$$\text{weight}(CB) + \text{weight}(DA) = 6 + 2 = 8,$$

whereas

$$\text{weight}(CD) + \text{weight}(BA) = 4 + 5 = 9.$$

So swap *D* and *B*. This gives the better cycle *ACBDA*, with weight 16.

Miscellaneous exercise 6

1 Consider the travelling salesperson problem for the network in the diagram.

(a) Apply the Lower Bound algorithm with *A*, *B*, *C* or *D* as the special node. What do you notice?

(b) Apply the Lower Bound algorithm with *E* as the special node.

(c) Explain why the bound of part (b) cannot be attained.

(d) What is the optimum solution?

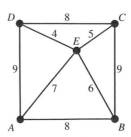

2 The distances shown on the network are in kilometres.

(a) Apply the Nearest Neighbour algorithm from Birmingham.

(b) Improve the tour of part (a) by using the Tour Improvement algorithm.

(c) Apply the Lower Bound algorithm with Leeds as the special node. What can you now say about the optimum solution to the travelling salesperson problem for this network?

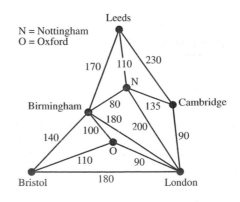

3 A firm uses certain specified routes between the following cities for its deliveries. The distances are in miles.

	Bi	Br	Ld	Le	Li	Lo	M	N	Sh	So
Birmingham	–	138	–	68	157	190	–	95	–	–
Bristol	138	–	–	–	–	193	–	–	–	122
Leeds	–	–	–	–	115	–	67	–	54	–
Leicester	68	–	–	–	–	150	–	40	–	–
Liverpool	157	–	115	–	–	–	61	173	–	–
London	190	193	–	150	–	–	–	–	–	123
Manchester	–	–	67	–	61	–	–	100	61	–
Nottingham	95	–	–	40	173	–	100	–	72	–
Sheffield	–	–	54	–	–	–	61	72	–	–
Southampton	–	122	–	–	–	123	–	–	–	–

(a) Apply the Nearest Neighbour algorithm starting from Leicester.

(b) Find the weight of a minimum spanning tree for the network with Southampton deleted.

(c) Make a deduction about the optimum solution to the travelling salesperson problem.

4 Draw three simple networks, with Hamiltonian paths, such that the optimum solutions to the Travelling Salesperson problem have weights

(a) less than twice,   (b) exactly twice,   (c) more than twice

the weight of the minimum connector.

5 A ring main consists of a loop of cable running from the electricity meter M to five double sockets in turn and then back to the meter. The estimated cost of each possible stretch of cable is as shown in the table.

	M	A	B	C	D	E
M	–	13	14	13	16	12
A	13	–	12	–	–	11
B	14	12	–	16	–	–
C	13	–	16	–	14	–
D	16	–	–	14	–	13
E	12	11	–	–	13	–

(a) Apply the Nearest Neighbour algorithm from M, then A, and finally E.

(b) Apply the Lower Bound algorithm from M. What can you deduce about the least cost of laying the ring main cable?

(c) Fitting each double socket costs £25. What effect does this have on the optimum path for laying the cable?

6 A modification of the Nearest Neighbour algorithm for use with the *practical* travelling salesperson problem is as follows.

Step 1 Choose any starting node.

Step 2 Choose a node which is at a minimal distance from the previously chosen mode.

Step 3 Repeat Step 2 until all nodes have been chosen.

Apply this algorithm to Question 5(a), starting from A.

7 An assembly line is used to produce five items, A, B, C, D and E. The times (in minutes) needed for each possible changeover are shown in the table.

	To A	B	C	D	E
From A	–	30	50	70	40
B	20	–	30	80	50
C	60	50	–	20	30
D	40	70	40	–	40
E	80	20	30	40	–

All five items need to be produced, and the assembly line must be returned to its initial state. The factory manager wishes to minimise the changeover time.

(a) What is the result of applying the Nearest Neighbour algorithm from A?

(b) Apply the Lower Bound algorithm with A as the special node.

(c) Improve the upper bound of part (a).

8 Find a Hamiltonian cycle in this graph of the dodecahedron. (This was the problem initially considered by Hamilton.)

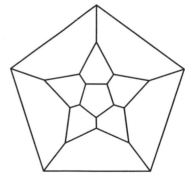

9 Find a Hamiltonian cycle for all $n \times n$ rectangular arrays of nodes with n even. The case $n = 4$ is illustrated.

10* Prove that an $n \times n$ rectangular array of nodes, with n odd, cannot have a Hamiltonian cycle.

11 Pirate Pete is hunting for buried treasure. He knows that the treasure is buried at one of the places marked on his map, shown in the figure, but he cannot solve the clues to find out where he should dig.

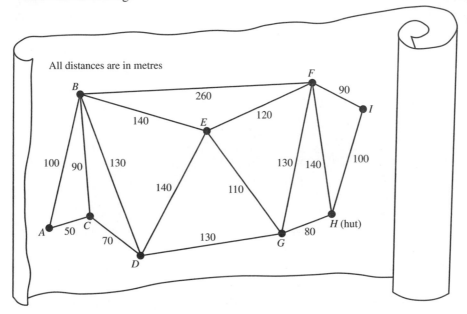

Pirate Pete decides to dig at each of the places marked on his map.

(a) Demonstrate, clearly, the use of a greedy algorithm, starting from the hut, H, to construct a minimum spanning tree (minimum connector) for the network.

(b) Explain, briefly, why twice the length of the minimum spanning tree gives an upper bound to the length of the route which is the solution to the practical travelling salesperson problem for the network.

(c) Write down a route that gives a better upper bound than that found in part (b), and state the length of this route.   (OCR)

12 Dina Mite has set up a fireworks display involving eight different fireworks, A, B, ... ,
    H. She had wanted to ignite the eight fireworks simultaneously. However, due to a lack of
    fuse wire, she has to sprint from one firework to another until she has ignited all eight, and
    then sprint back to safety.

    It takes Dina 10 seconds to sprint from safety to any one of the fireworks, or to sprint from
    a firework back to safety. It takes her 2 seconds to ignite each firework, and the times in
    seconds taken to sprint between the fireworks are given in the table.

	A	B	C	D	E	F	G	H
A		2	9	7	9	6	9	6
B	2		8	5	8	5	8	4
C	9	8		2	5	6	8	8
D	7	5	2		4	5	6	2
E	9	8	5	4		3	4	6
F	6	5	6	5	3		3	5
G	9	8	8	6	4	3		5
H	6	4	8	2	6	5	5	

(a) Use Prim's algorithm, starting from A, to find the minimum connector for the eight
     fireworks. Draw a diagram showing the minimum connector, and write down its total
     length, T.

Let θ be the time that it takes Dina to get from safety, to ignite all eight fireworks and
return to safety.

(b) Explain why $\theta \geqslant 36 + T$.

(c) Explain why $\theta \leqslant 36 + 2T$.

Dina has 45 seconds from when she has ignited the first firework to when she reaches
safety.

(d) Find a route that enables Dina to ignite all eight fireworks and get back to safety in
     time.   (OCR)

7 Linear programming

In this chapter you will learn how to tackle a range of maximising and minimising problems subject to various conditions or constraints. When you have completed it you should

- know what objective functions and linear constraints are
- be able to formulate practical problems as linear programming problems
- be able to solve linear programming problems by graphical means.

7.1 Optimisation

In an optimisation problem, the objective is to optimise (maximise or minimise) some function. Typical problems from the world of business might include

- maximising profits
- minimising costs
- maximising turnover
- minimising the time needed
- maximising the number of customers.

The following example is somewhat simplified, but it gives you an idea of the kind of situation which can arise.

As an example of a linear programming problem and its formulation as a problem, consider a recycling plant which produces two types of paper, P_1 and P_2, by using a mixture of scrap paper and timber.

Each tonne of paper P_1 requires 2 tonnes of scrap paper, and each tonne of paper P_2 requires 1 tonne of scrap paper, and there is a maximum of 16 tonnes of scrap paper available.

In addition, each tonne of paper P_1 requires 2 tonnes of timber, and each tonne of paper P_2 requires 3 tonnes of timber, and there is a maximum of 24 tonnes of timber available.

How can the plant maximise the total amount of paper produced per day?

You may find it helpful to summarise the information in a table like Table 7.1.

Raw material	Raw material per tonne of paper		Availability per day
	P_1	P_2	
Scrap paper	2 tonnes	1 tonne	16 tonnes
Timber	2 tonnes	3 tonnes	24 tonnes

Table 7.1

First you need to define the variables in terms of which the problem can be formulated. In this case, suppose the plant produces x tonnes of P_1 and y tonnes of P_2 each day.

As x tonnes of P_1 and y tonnes of P_2 requires $2x + y$ tonnes of scrap paper, and, as only 16 tonnes are available, $2x + y \leqslant 16$.

Similarly x tonnes of P_1 and y tonnes of P_2 requires $2x + 3y$ tonnes of timber, and only 24 tonnes are available. Therefore $2x + 3y \leqslant 24$.

In addition, you cannot produce negative amounts of P_1 and P_2, so $x \geqslant 0$ and $y \geqslant 0$.

The inequalities $x \geqslant 0$ and $y \geqslant 0$ are likely to be part of every problem like this.

The total amount of paper produced per day, which has to be maximised, is $x + y$.

Summarising this,

maximise $\qquad x + y$,

subject to $\qquad 2x + y \leqslant 16,$ $\qquad$ (scrap paper)
$\qquad\qquad\quad\; 2x + 3y \leqslant 24,$ $\qquad$ (timber)
$\qquad\qquad\quad\; x \geqslant 0,$
$\qquad\qquad\quad\; y \geqslant 0.$

Now that the problem has been stated, it is time to take stock with some definitions.

The variables x and y used to formulate the problem are called **control variables**. The function to be optimised is called the **objective function**. The inequalities are examples of **constraints**; for reasons which will become clear later, they are called 'linear' constraints. So, in this problem you have to maximise the objective function $x + y$ subject to the four linear constraints.

The solution to the problem will be continued in Section 7.3, after some important graphical ideas in Section 7.2. You may wish to think about its solution before going further.

7.2 Representing inequalities graphically

In P1 Section 6.1 you saw that you could visualise an inequality such as $a > b$ by saying that a lies to the right of b on a number line (see Fig. 7.2).

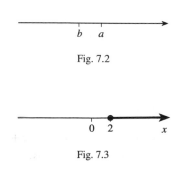

Fig. 7.2

Fig. 7.3

You can describe the inequality $x \geqslant 2$ by saying it is satisfied by all the points on the number line which lie to the right of the point 2, or at the point 2 itself. It can be represented by a diagram such as Fig. 7.3, where the solid blob shows that the point $x = 2$ is included.

Now look at the inequality $y \geqslant x+1$. You know that $y = x+1$ is the equation of a line, and that $y = x+1$ is also a rule, P1 Section 1.4, for determining whether or not a given point in the plane with coordinates (x, y) lies on the line. Fig. 7.4 shows the graph of $y = x+1$.

If the equation is not satisfied by a given point, then the point does not lie on the line, so it is either above the line or below the line.

For such a point $y \neq x+1$, so either $y > x+1$ or $y < x+1$. The points which satisfy $y > x+1$ or $y < x+1$ are the points which are not on the line.

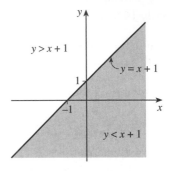

Fig. 7.4

Points above the line (the unshaded region) have coordinates (x, y) which satisfy $y > x+1$. The points below the line satisfy $y < x+1$. For example, the point $(2, 4)$ has the property that $y > x+1$, and it is clearly above the line.

> In general, the line $y = mx + c$ divides the plane into three regions.
>
> - The points on the line satisfy $y = mx + c$.
> - Points in the region above the line satisfy $y > mx + c$.
> - Points in the region below the line satisfy $y < mx + c$.

The inequality $y \geqslant x+1$ consists of the points which satisfy the line $y = x+1$ and the inequality $y > x+1$. It therefore consists of the line in Fig. 7.4, together with the unshaded points above the line.

Note that to illustrate $y > x+1$ it is conventional to draw $y = x+1$ as a dotted or dashed line, and to shade the region you do not require.

Example 7.2.1

Draw diagrams to show the regions satisfied by the inequalities
(a) $y \geqslant x$, (b) $2x + y \leqslant 2$, (c) $x \geqslant 0$, (d) $x - 2y \leqslant 1$.
In each case, shade the unwanted region.

The diagrams are shown in Fig. 7.5, after the commentary.

(a) Draw the line $y = x$. Then the required region consists of the points on the line, and the points which lie above the line. Notice that the shading is on the other side of the line, and rules out the points which do *not* satisfy the inequality. It is conventional to indicate the required region by shading in this way; it is also usual not to shade out the whole of the unwanted region, as otherwise these diagrams tend to become messy.

(b) Since you can write $2x + y \leqslant 2$ in the form $y \leqslant 2 - 2x$, you can now use the method in the shaded box.

(c) Strictly, the shaded box does not apply to inequalities of this type, but it is easy to see that $x \geqslant 0$ corresponds to the region to the right of $x = 0$.

(d) You need to take some care with this because of the sign of y. Since
$x - 2y \leq 1$, $x - 1 \leq 2y$, so $y \geq \frac{1}{2}x - \frac{1}{2}$.

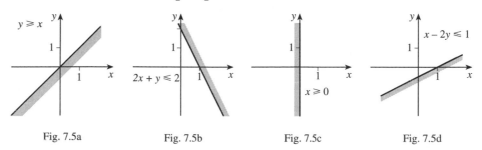

Fig. 7.5a   Fig. 7.5b   Fig. 7.5c   Fig. 7.5d

Example 7.2.2

Indicate by shading the region whose points satisfy the constraints $y \geq x$, $2x + y \leq 2$,
$x \geq 0$ and $x - 2y \leq 1$. Use your diagram to find the coordinates of the
point in the region which has the greatest x-coordinate.

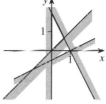

These constraints are just the ones of Example 7.2.1.
So put the diagrams in Fig. 7.5 together, as in
Fig. 7.6. The required region is the small triangle
with no shading in it. The point with the greatest
x-coordinate is found to be $\left(\frac{2}{3}, \frac{2}{3}\right)$ by solving the
simultaneous equations $2x + y = 2$ and $x = y$.

Fig. 7.6

Notice that the constraint $x - 2y \leq 1$ plays no part in determining the region that
satisfies all four constraints.

*If in this example you had shaded the wanted regions, then the required region (the small
triangle) would have been shaded four times, and it would be harder to pick out.*

Notice also that in Example 7.2.2, you were effectively asked to maximise the value of
x over all the points in that region. The maximum value of x turned out to be $\frac{2}{3}$.

Now suppose that you had been asked to maximise
the value of $x + y$ in the triangular region not shaded
in Fig. 7.6. In Fig. 7.7 you will see a magnified
version of this region and a number of parallel lines,
all of the form $x + y = k$ for various values of k. The
value of k for each line is shown either at the bottom
of the graph or along the right side.

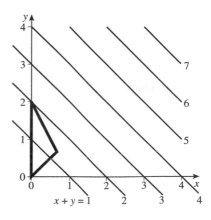

The important thing to notice is that the value of k
increases as the parallel lines move to the right. The
maximum value of $x + y$ for the region is the line
furthest to the right which passes through a point of
the region. This is the line which passes through
$(0, 2)$, giving $x + y$ equal to $2 + 0 = 2$.

Fig. 7.7

Example 7.2.3

Maximise $2x + y$ subject to the constraints $3x + y \leqslant 6$, $x + 2y \leqslant 7$, $x \geqslant 0$ and $y \geqslant 0$.

Fig. 7.8 shows the region which satisfies the four constraints. In addition, a line parallel to the family of lines $2x + y = k$ has been drawn. As k increases, these parallel lines get further to the right (and as it decreases they get further left), so the maximum value of $2x + y$ arises when the line passes through the marked point, which has coordinates $(1, 3)$. This maximum value is then $2 \times 1 + 3 = 5$.

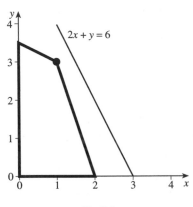

Fig. 7.8

Note that you find graphically which point gives the maximum, and then calculate its coordinates by solving the simultaneous equations $3x + y = 6$ and $x + 2y = 7$.

This situation is characteristic of the general case. As there are only two control variables, the feasible region can be illustrated well by a two-dimensional drawing. Suppose that the region formed by the constraints is bounded by straight lines, and is convex (that is, the corners all point outwards). Then, if the objective function to be maximised is of the form $ax + by$, where a and b are constants, its value is maximised at one of the vertices of the region.

In the special case when one of the edges of the region is parallel to the direction of the objective function line, the objective function may be maximised at two of the vertices, and at every point joining those two vertices. Exercise 7A Question 5 gives an example of this.

<hr>

Exercise 7A

1   In each part, sketch the regions determined by the inequalities. Find the maximum value of x and the maximum value of y in each of the regions. Keep your sketches for Question 2.

   (a) $x + 3y \leqslant 6$, $x \leqslant 3$, $x \geqslant 0$, $y \geqslant 0$

   (b) $x + y \leqslant 4$, $x - y \leqslant 0$, $x \geqslant 0$, $y \leqslant 3$

   (c) $-x + y \geqslant -3$, $-x + y \leqslant 3$, $y \leqslant 4$, $x \geqslant 0$, $y \geqslant 0$

   (d) $x - 4y \leqslant 4$, $x + y \leqslant 5$, $x \geqslant 0$, $y \geqslant 0$

2   Maximise the objective function $x + 2y$ for each set of constraints in Question 1.

3   Maximise the function $y - x$ subject to the constraints $2x - y \geqslant -3$, $x - 2y \leqslant 3$, $2x + y \leqslant 11$, $x \geqslant 0$ and $y \geqslant 0$.

4   Minimise the function $y - x$ subject to the constraints in Question 3.

5   Maximise $3x + y$ subject to the constraints $2y \geqslant 3x$, $y + 3x \leqslant 9$, $x \geqslant 0$ and $y \geqslant 0$.

7.3   Linear programming problems

Here is the summary of the problem in Section 7.1, about making two different kinds of paper.

Maximise $x + y$,

subject to $2x + y \leqslant 16$,   (scrap paper)
$\quad 2x + 3y \leqslant 24$,   (timber)
$\qquad\quad x \geqslant 0$,
$\qquad\quad y \geqslant 0$.

The solution is shown in Fig. 7.9. The region which contains the allowable points has been highlighted, and the line $x + y = 16$ has been drawn with a dashed line. You can see that the point in the region which maximises $x + y$ is the point $(6, 4)$. The maximum value itself is therefore $6 + 4$, which is 10.

The maximum amount of paper which can be made per day is 10 tonnes.

In general, the region which contains the allowable points is called the **feasible region**. The variables x and y which describe the problem are called **control variables**. The constraints are called 'linear' constraints because they are closely related to straight lines.

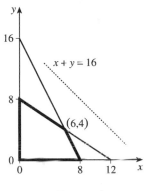

Fig. 7.9

Since World War II there have been major advances in the techniques for dealing with optimisation problems that have large numbers of linear constraints. In particular, computer software which can solve a wide range of practical problems is now readily available. In this chapter, you will learn a method for dealing with large-scale problems, but first you will gain experience with small-scale problems which can be solved graphically. This branch of mathematics is called **linear programming**, because of the connection with straight lines.

In general, for variables x, y, z, ... , a combination of these variables is called **linear** if it is of the form

$ax + by + cz + \ldots$, where a, b, c, ... are constants.

Example 7.3.1
Two machines, M_1 and M_2, are used to make two types of lamp, L_1 and L_2. Lamp L_1 requires the use of machine M_1 for 2 minutes and machine M_2 for 3 minutes. Lamp L_2 requires the use of machine M_1 for 4 minutes and machine M_2 for 3 minutes. The profit on lamp L_1 is £7 and the profit on lamp L_2 is £11. How can the profit per hour be maximised?

Define the control variables x and y to be the number of lamps L_1 and L_2 produced each hour. Then the profit per hour is $£(7x+11y)$, so $7x+11y$ needs to be maximised.

It is useful to lay out the information in tabular form.

Lamp	Time on M_1	Time on M_2
L_1	2 minutes	3 minutes
L_2	4 minutes	3 minutes
Time available	60 minutes	60 minutes

The constraints, apart from $x \geqslant 0$ and $y \geqslant 0$, are $2x+4y \leqslant 60$, arising from the use of M_1, and $3x+3y \leqslant 60$, arising from the use of M_2.

Summarising, and dividing the constraint inequalities by 2 and 3 respectively, the problem becomes:

$$\text{maximise} \quad P = 7x+11y,$$

$$\text{subject to} \quad \begin{aligned} x+2y &\leqslant 30, &&(\text{machine } M_1) \\ x+y &\leqslant 20, &&(\text{machine } M_2) \\ x &\geqslant 0, \\ y &\geqslant 0. \end{aligned}$$

The solution is shown in Fig. 7.10. The region which contains the allowable points has been highlighted, and the line $7x+11y=231$ has been drawn to show the direction of $7x+11y=k$. You can see that the point in the region which maximises $7x+11y$ is the one marked in the diagram. By solving the equations $x+2y=30$ and $x+y=20$ simultaneously, you find that the point is $(10,10)$. The maximum value itself is therefore $7 \times 10 + 11 \times 10$, which is 180.

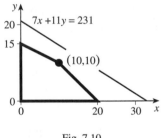

Fig. 7.10

The profit is maximised by making 10 of each lamp each hour.

An alternative method is to draw the graph to find all the vertices of the feasible region. Since you know that the objective function is maximised or minimised by the coordinates of a vertex, you can simply test each vertex in turn to find the best one. This method is used in the next example.

Example 7.3.2

A joiner makes two kinds of bookcase, standard and de luxe. The standard bookcase requires 4 square metres to make and store it, and the de luxe model requires 5 square metres, and there are only 61 square metres of space available. The standard model takes 30 minutes to make, and the de luxe model takes 40 minutes, and there are 480 minutes available in the day. The profit on a standard model is £50, and on a de luxe model it is £70. How should the joiner's time be spent?

The problem may be summarised as follows.

Maximise $50x + 70y$,

subject to
$$4x + 5y \leqslant 61, \quad \text{(available space)}$$
$$30x + 40y \leqslant 480, \quad \text{(available time)}$$
$$x \geqslant 0, \ y \geqslant 0.$$

From a graph, the coordinates of the feasible region are found to be $(0,0)$, $\left(15\frac{1}{4},0\right)$, $(4,9)$ and $(0,12)$.

The values of the objective function at these vertices in turn are 0, $762\frac{1}{2}$, 830 and 840. Since the largest value, 840, corresponds to the point $(0,12)$, the joiner should make only de luxe bookcases.

Note that the point $\left(15\frac{1}{4},0\right)$ does not give a practical solution. This is covered in Section 7.5.

7.4 Blending problems

Blending is an important and highly mathematical commercial activity, whether it be blending fruit juices, paint or margarine oils. Formulating constraints mathematically for blending problems can be quite complicated. The solution to Example 7.4.1 provides a useful illustration of the standard techniques.

Example 7.4.1

A blending company buys two types of fruit juice, A and B, from other suppliers. The details are summarised in the table below.

Juice	Orange juice	Lemon juice	Other	Cost per litre	Minimum weekly order
A	50%	0%	50%	40p	25 000 litres
B	20%	10%	70%	30p	30 000 litres

These juices are blended to produce a fruit juice which must contain at least 30% orange juice and at least 5% lemon juice. How can the company minimise the cost of producing at least 60 000 litres of juice per week?

Suppose the company uses x litres of A and y litres of B.

Since the new fruit juice has to contain 30% orange juice,

$$\frac{0.5x + 0.2y}{x+y} \geqslant 0.3,$$

which simplifies to give $2x \geqslant y$.

Similarly, the constraint on lemon juice gives

$$\frac{0.1y}{x+y} \geqslant 0.05,$$

which can be rearranged to give $y \geqslant x$.

The other constraints are the total requirement, which is $x + y \geqslant 60\,000$, and the minimum orders, which are $x \geqslant 25\,000$ and $y \geqslant 30\,000$.

Here is the summary of the problem.

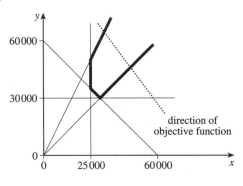

Fig. 7.11

Minimise $P = 0.4x + 0.3y,$

subject to $2x \geqslant y,$
                   $y \geqslant x,$
                   $x + y \geqslant 60\,000,$
                   $x \geqslant 25\,000,$
                   $y \geqslant 30\,000.$

From Fig. 7.11, the minimum occurs when $x = 25\,000$ and $y = 35\,000$, giving a cost of £20,500.

The next section deals with a complication which can arise in some linear programming problems.

7.5 Integer solutions

Consider the recycling plant example used in Section 7.1, and suppose that the availability of scrap paper and timber dropped to 13 and 21 tonnes per day respectively. The information is laid out in Table 7.12.

Raw material	Raw material per tonne of paper		Availability per day
	P_1	P_2	
Scrap paper	2 tonnes	1 tonne	13 tonnes
Timber	2 tonnes	3 tonnes	21 tonnes

Table 7.12

As a linear programming problem, this would be

maximise $x + y,$

subject to $2x + y \leqslant 13,$
                $2x + 3y \leqslant 21,$
                $x \geqslant 0, y \geqslant 0.$

If you work through the problem you will find that the maximum value of $x + y$ is 8.5, obtained when 4.5 tonnes of P_1 and 4 tonnes of P_2 are produced. In some contexts this output of 8.5 tonnes would be the required solution. However, there are circumstances where only integer solutions are appropriate. In this case, for example, the various contracts for the recycling plant's paper might be for whole numbers of tonnes. In that case, the linear programming problem is best formulated as follows.

Maximise $x + y$,

subject to $2x + y \leqslant 13$,
$2x + 3y \leqslant 21$,
$x \geqslant 0, \ y \geqslant 0$,
x, y are integers.

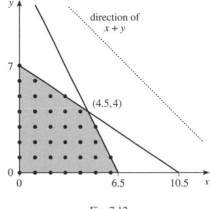

The integer points inside the feasible region are shown in Fig 7.12. The solution is 8 tonnes, obtained at three different points $(3,5)$, $(4,4)$ and $(5,3)$. Note that two of these are on the boundary of the feasible region, but not at a node, and $(4,4)$ is not even on the boundary.

Fig. 7.12

In some cases, the integer solution may be some distance from the node at which the 'normal' solution occurs. See Exercise 7B Question 5.

Exercise 7B

1 A company has 2400 assembly hours available for a trial run of two new models of calculator. The availability of components means that at most 400 of each can be manufactured initially. The assembly times and profits for the two models are as shown in the table. The total profit has to be maximised.

Model	Assembly hours	Profit
A	3	£7
B	4	£8

(a) Formulate this as a problem with integer solutions.

(b) Use a graphical method to maximise the total profit.

2 A small factory makes two types of valve, V and W, for profits of £5 and £3 respectively. Let x be the number of V and y be the number of W produced per day. Maximise the total daily profit subject to the constraints

$x + 2y \leqslant 90$,   (machine hours)

$2x + y \leqslant 60$.   (labour)

3 Due to market changes, the profit on each valve W of Question 2 changes to £p. For what values of p should the factory

(a) make only valve V,   (b) make only valve W?

4 A factory produces two items. Each day there are 160 labour hours available and 200 machine hours. Item 1 requires 2 hours of labour and 5 hours of machine time. Item 2 requires 4 hours of labour and 2 hours of machine time.

(a) Formulate the problem of maximising the total daily output of items as a linear programming problem.

(b) How should the production be divided between the two items?

(c) Suppose the profit is £50 on item 1 and £10 on item 2. How would this affect your answer to part (b)?

5 Solve these two linear programming problems.

(a) Maximise $2x + 3y$,

subject to $20x + 32y \leqslant 160$,
                $3x - 2y \leqslant 0$,
                $x \geqslant 0, y \geqslant 0$.

(b) Maximise $2x + 3y$,

subject to $20x + 32y \leqslant 160$,
                $3x - 2y \leqslant 0$,
                $x \geqslant 0, y \geqslant 0$,
                x, y are integers.

6 A brand of margarine is made from two kinds of oil, which are refined and then blended.

Oil	Cost per kg	Hardness	Maximum daily production
X	£1.50	3.2	5000 kg
Y	£1.80	9.5	3000 kg

The margarine sells at £3.50 per kg and its hardness must be between 6 and 7.

(a) Formulate this as a linear programming problem to maximise the daily profit.

(b) How can the manufacturers maximise their daily profit?

7 Expected annual returns are 4% from savings accounts and 10% from shares. An investor requires an annual return of at least 7.5% and has up to £10,000 to invest. The investor is 'risk averse' and so wants to invest as much as possible in a savings account.

(a) Formulate this as a linear programming problem.

(b) Hence advise the investor.

8 The budget for a promotion run by a small company is £10,000. Full-page advertisements in regional newspapers cost £1200 each and have an estimated audience of 40 000 people. Advertisements on radio programmes cost £500 each and have an estimated audience of 18 000 people. It has been decided that at most ten of the advertisements will be on the radio. Assuming no overlaps between the various audiences, how many advertisements should be scheduled in each medium to maximise audience contact? What is the total cost and the total audience?

9 A company produces margarines by blending three oils, A, B and C. Here are the details for the next production run.

Oil	Cost per kg (£)	Availability (kg)
A	1.25	5000
B	1.60	10 000
C	1.82	10 000

Two brands of margarine are made, Regular and De luxe, selling at £2.50 per kilogram and £3.50 per kilogram respectively. Regular must consist of at most 30% oil A and at least 40% oil B. De luxe must consist of at most 40% oil B and at least 30% oil C. Formulate, but do not solve, the problem of maximising the profit by letting x, y and z be the number of kilograms of A, B and C, respectively, used for Regular and by letting u, v and w be the number of kilograms of A, B and C, respectively, used for De luxe.

Miscellaneous exercise 7

1   Gin is 45% alcohol by volume and tonic is non-alcoholic. It is required to mix gin-and-tonics which are at least 10% alcohol by volume and contain at least 200 ml of liquid but at most 30 ml of alcohol.

   (a) What are the least and greatest amounts of gin which such a drink could contain?

   (b) What are the least and the greatest amounts of tonic which such a drink could contain?

2   Solve the two linear programming problems:

   (a) Maximise $2x + 3y$,   (b) Maximise $2x + 3y$,

        subject to $5x + 7y \leqslant 35$,   subject to $5x + 7y \leqslant 35$,

                            $4x + 9y \leqslant 36$, $4x + 9y \leqslant 36$,

                            $x \geqslant 0, y \geqslant 0$. $x \geqslant 0, y \geqslant 0$,

                                                                                    x, y are integers.

3   The following figures can be used to compare the three main forms of investment.

	Deposits	Gilts	Equities
Expected return[1]	1.5%	2.5%	8%
Risk factor[2]	0	20	60

   [1]   Real return is inflation adjusted. Based upon the average return over the last 80 years.
         (Source: Barclays Capital Equity Gilt Study 1999)

   [2]   Based upon the expected volatility of the invested capital.

   A retired person has £70,000 to invest to supplement their pension. A stockbroker recommends that they should have a mix of investments to give an expected return of at least 5.5% and an overall risk factor of at most 40. Subject to these constraints, the pensioner wishes to maximise the amount of money in fixed interest investment (gilts).

   (a) Let £x, £y and £z be the amounts invested in deposits, gilts and equities respectively. Show that the problem can be formulated as follows.

        Maximise y,

        subject to $x + y + z = 70\,000$, $8x + 6y \leqslant 5z$,

                        $2x + y \geqslant z$, $x \geqslant 0, y \geqslant 0, z \geqslant 0$.

   (b) What main assumption has been made in this formulation of an investment problem?

   (c) Solve the linear programming problem of part (a). (Hint: substitute for z.)

4   (a) Sketch the feasible region for the following constraints.

        $-7x + 10y \leqslant 70$ $x + y \leqslant 15$ $x \leqslant 7$

        $7x - 4y \leqslant 28$ $21x + 7y \geqslant 63$ $5x + 11y \geqslant 55$

        $x, y \geqslant 0$.

   (b) What value of x maximises

        (i) $4x + 5y$,   (ii) $5x + 4y$?

5   A university has enough places for 5000 students. Government restrictions mean that at least 80% of the places must be given to UK students, but the remainder may be given to overseas students.

There are 2000 residential places available in the halls. All overseas students and at least one-third of the UK students must be given places in the halls.

The university gets £4000 in tuition fees for each UK student and £6000 for each overseas student. It wants to maximise the fees received.

Let x be the number of places given to UK students and y be the number of places given to overseas students.

(a) Explain, briefly, why the problem requires the function $P = 4000x + 6000y$ to be maximised.

(b) The constraints of the problem are
$$x + y \leqslant 5000,$$
$$4y \leqslant x,$$
$$\tfrac{1}{3}x + y \leqslant 2000,$$
$$x \geqslant 0 \text{ and } y \geqslant 0.$$

Explain why

(i) $4y \leqslant x$, and   (ii) $\tfrac{1}{3}x + y \leqslant 2000$.

The feasible region for the problem is shown in the figure.

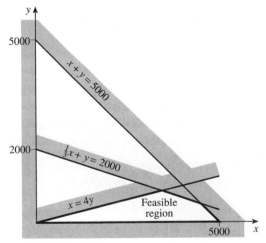

(c) Write down the vertices of the feasible region which lie on the x-axis. Use simultaneous equations to calculate the coordinates of the other vertices of the feasible region (to the nearest whole number), and hence calculate the value of P at each vertex.

(d) Use your answer to (c) to advise the university on how many places they should give to UK students, and how many to overseas students, to maximise the fees received.

(OCR)

6 A company manufactures two kinds of robot, the Brainy and the Superbrainy. After production, each robot is tested for its coordination and its logic. The company wants to maximise its profit from the sale of the robots. You may assume that every robot manufactured is sold. The table below shows the times required by these tests, the time available each week, and the profit per robot.

	Coordination test (hours)	Logic test (hours)	Profit per robot (£)
Brainy	6	3	2400
Superbrainy	3	4	3000
Time available (hours)	60	60	

This problem is being modelled using a linear programming formulation:

x = number of Brainy robots manufactured per week;

y = number of Superbrainy robots manufactured per week.

Maximise $P = 2400x + 3000y$.

(a) Write down the four constraints for this problem.

(b) Represent the constraints graphically, marking the feasible region clearly.

(c) Showing your method clearly, solve the linear programming problem. (OCR)

7 Lou Zitt has a budget of £2000 to spend on storage units for his office. The storage units must not cover more than 50 m^2 of floor space. Lou wants to maximise the storage capacity.

The three types of storage unit that he can choose from are shown below.

Type	Storage capacity (m^3)	Floor space covered (m^2)	Cost (£)
Antique pine units	2	1	100
Beech wood units	9	4	500
Cedar wood units	5	3	200

Suppose that Lou buys a antique pine units, b beech wood units and c cedar wood units.

(a) Write down two constraints that must be satisfied by a, b and c, other than $a \geqslant 0$, $b \geqslant 0$ and $c \geqslant 0$.

(b) Write down the objective function for this problem.

(c) Set up the problem as an LP formulation. ('LP' stands for 'linear programming'.) You are not expected to solve the problem.

(d) Identify which aspect of the original problem has been overlooked in the LP formulation.

Because of the shape of the office, Lou also needs to consider the widths of the units. The antique pine units are 1 metre wide, the beech wood units are 2 metres wide and the cedar wood units are 3 metres wide. The total width of the units cannot exceed 20 metres.

(e) Show how to incorporate this additional restriction into your LP formulation. (OCR)

8 Simon is preparing for his motorcycle test. The test has two components, a practical test and a theory test.

Simon estimates that each hour spent preparing for the practical test and each half-hour spent preparing for the theory test will increase his final test score by 5 points. He wants to score as many points as possible.

Simon also knows that he needs to increase his score on the practical test by at least 10 points, that he cannot spend more than 9 hours preparing for the test (of which no more than 4 hours can be spent on preparing for the practical test), and that he must spend longer on preparing for the theory test than on preparing for the practical test.

(a) Identify the variables for this problem.

(b) Set up the problem as an LP formulation, with an objective function and a set of constraints that must be satisfied.   (OCR)

9 The owner of a sweet shop has a stock of 200 acid drops, 240 bubble gums and 240 candy bars. He uses these to make up two types of mixed bag.

Each bag of type X contains 4 acid drops, 6 bubble gums and 3 candy bars.
Each bag of type Y contains 4 acid drops, 3 bubble gums and 6 candy bars.

The profit on each bag of type X sold is 20 pence, and the profit on each bag of type Y sold is 25 pence.

The owner of the sweet shop wants to maximise the profit on the sales of these bags.

(a) Identify the variables for this problem.

(b) Write down three constraints that must be satisfied by the variables, other than that they must be non-negative integers.

(c) Write down the objective function for this problem.

(d) State an assumption that has been made in writing down the objective function.

(e) Represent the three constraints graphically, and hence find the maximum profit. You should explain your method carefully.

The owner of the sweet shop decides that there should be no bubble gums left over when all the bags have been sold.

(f) Explain how this additional restriction leads to a change in the point that gives the maximum profit, and state the maximum profit in this case.   (OCR)

8 The Simplex algorithm

This chapter is about solving linear programming problems when graphical methods are not easily available. When you have completed it you should

- be able to solve linear programming problems by the Simplex algorithm.

8.1 The Simplex method

The graphical method you have used in Sections 7.2, 7.3 and 7.4 is an excellent method when it works. However, it is useful to be able to solve problems without drawing a diagram. The Simplex method has been designed to do just this. Such a method is of course essential for solving problems on a computer. It is also essential if you have a practical problem with lots of variables which cannot satisfactorily be represented by a two-dimensional drawing.

Consider the opening question of Sections 7.1 and 7.3, which was

$$\text{maximise} \quad P - x + y,$$

$$\text{subject to} \quad 2x + y \leqslant 16,$$
$$2x + 3y \leqslant 24,$$
$$x \geqslant 0, y \geqslant 0.$$

To use the Simplex method, you first introduce **slack variables**, s and t say, to convert the two non-trivial inequalities into equalities.

Let $\quad s = 16 - 2x - y \qquad\qquad$ Equation 1

and $\quad t = 24 - 2x - 3y. \qquad\qquad$ Equation 2

Then the inequalities reduce to $x \geqslant 0, y \geqslant 0, s \geqslant 0, t \geqslant 0$.

The vertices of the feasible region (see Fig. 8.1) then correspond to places where two of the four variables are zero. Check, for example, that A is the point where $y = 0$ and $s = 0$.

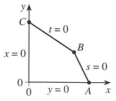

Fig. 8.1

The points O, A, B and C are important because, as you have seen, the objective function will be maximised at one of them.

Equations 1 and 2 can be used to write the objective function P in terms of any pair of variables. For example, writing x as $P - y$ and substituting in Equation 1 you get

$$s = 16 - 2(P - y) - y, \quad \text{which reduces to} \quad s = 16 - 2P + 2y - y = 16 - 2P + y.$$

Making P the subject of the equation $s = 16 - 2P + y$ gives

$$P = 8 + \tfrac{1}{2}y - \tfrac{1}{2}s.$$

You should check that you can obtain the alternatives

$$P = x + y, \quad P = 8 + \tfrac{1}{2}y - \tfrac{1}{2}s, \quad P = 10 - \tfrac{1}{4}s - \tfrac{1}{4}t, \quad P = 8 + \tfrac{1}{3}x - \tfrac{1}{3}t.$$

The third expression, $P = 10 - \tfrac{1}{4}s - \tfrac{1}{4}t$, is special because the coefficients of the two variables, s and t, are both negative. Since $s \geqslant 0$ and $t \geqslant 0$ this means that P *cannot* be greater than 10. The maximum is therefore 10, occurring when $s = t = 0$, which is at the point where $x = 6$ and $y = 4$.

This method seems much clumsier than the graphical method! However, it can be speeded up by adopting a clever notation for the linear equations which need to be manipulated to find the expressions for P. Moreover, it can be programmed for many variables.

8.2 The tableau format

Consider solving the two linear equations

$$x + 2y = 7, \qquad\qquad \text{Equation 1}$$
$$3x - 4y = -9. \qquad\qquad \text{Equation 2}$$

It is legitimate to double an equation:

$$2x + 4y = 14. \qquad\qquad \text{Equation 3} = 2 \times \text{Equation 1}.$$

At this stage it is useful to have a shorter notation for Equation 1 and Equation 2. A number in a circle, such as ①, will mean Equation 1. Thus the previous line, which was explained by Equation 3 = 2 × Equation 1, would now be explained by ③ = 2 × ①.

Returning to sets of equations, it is also legitimate to add (or subtract) them.

Thus adding ② and ③ gives

$$5x = 5. \qquad\qquad ④ = ② + ③$$

Dividing this equation by 5, you get

$$x = 1. \qquad\qquad ⑤ = \tfrac{1}{5} \times ④$$

Continuing, you should obtain $y = 3$.

You can write these operations neatly in a tableau, shown in Table 8.1.

x	y		Equation	
1	2	7	①	
3	−4	−9	②	
2	4	14	③ = 2 × ①	
5	0	5	④ = ② + ③	
1	0	1	⑤ = $\tfrac{1}{5}$ × ④	This equation is just $x = 1$.
0	2	6	⑥ = ① − ⑤	
0	1	3	⑦ = $\tfrac{1}{2}$ × ⑥	This equation is $y = 3$.

Table 8.1

The tableau format becomes even more useful as the number of variables increases. Consider, for example, the manipulation of equations which occurred in Section 8.1.

Example 8.2.1
Given that $P = x + y$, $2x + y + s = 16$ and $2x + 3y + t = 24$, use a tableau to express P in terms of s and t.

P	x	y	s	t			Equation
1	−1	−1	0	0	0		①
0	2	1	1	0	16		②
0	2	3	0	1	24		③
1	0	$-\frac{1}{2}$	$\frac{1}{2}$	0	8		④ = ① + $\frac{1}{2}$×②
0	0	2	−1	1	8		⑤ = ③ − ②
1	0	0	$\frac{1}{4}$	$\frac{1}{4}$	10		⑥ = ④ + $\frac{1}{4}$×⑤

Equation ② is used to eliminate x. Then y is eliminated.

So $P + \frac{1}{4}s + \frac{1}{4}t = 10$, giving $P = 10 - \frac{1}{4}s - \frac{1}{4}t$, as before.

In the tableau format, rows can be

• multiplied by constants or divided by (non-zero) constants,

• added or subtracted.

8.3 The Simplex algorithm

The ideas of Sections 8.1 and 8.2 can now be combined to give you a general (and easily programmed) method of solving a linear programming **maximising** problem.

The objective function must be expressed using an equation in which the right side is a number. For example, if $P = 7x + 11y$, then this is written as $P - 7x - 11y = 0$. Each non-trivial constraint must also be expressed as an equation by using slack variables. For example, $2x + 4y \leqslant 60$ would be written as $2x + 4y + s = 60$, where $s \geqslant 0$, and $2x + 4y \geqslant 60$ would be written as $2x + 4y - s = 60$, where $s \geqslant 0$.

The shaded box on the next page describes the Simplex algorithm, and the example after it illustrates how to do the algorithm in practice.

A commentary on the whole process follows after the tableau. You are advised to read the algorithm and study the tableau and commentary together.

Note that, for $2x + 4y + s = 60$, s is the 'slack' in the inequality $2x + 4y \leqslant 60$. For example, if $x = 6$ and $y = 10$, then $2x + 4y = 52$, and the slack is $s = 8$.

The Simplex algorithm

Step 1   Formulate the maximising problem (using slack variables as necessary) as a tableau.

Step 2   Ensure that all elements in the last column (except possibly the top one) are non-negative.

Step 3   Select any column (except the last one) whose top element is negative.

Step 4   Call the numbers in the selected column a_0, a_1, a_2, ... and the numbers in the last column l_0, l_1, l_2, ... In the selected column, choose the positive element a_i for which $\dfrac{l_i}{a_i}$ is least. This a_i is called the **pivot**.

Step 5   Divide the ith row by a_i.

Step 6   Combine appropriate multiples of the ith row with all the other rows in order to reduce to zero all other elements in the column of the pivot.

Step 7   If all top elements (except possibly the last one) are non-negative then the maximum has been reached. Otherwise return to Step 3.

Step 8   The last column contains the values of the objective function and the non-zero variables.

Example 8.3.1   (See Example 7.3.1)

Maximise $7x + 11y$ subject to the constraints $2x + 4y \le 60$, $3x + 3y \le 60$ and $x, y \ge 0$.

Start by introducing the slack variables s and t, by writing $2x + 4y + s = 60$ and $3x + 3y + t = 60$, so that $s \ge 0$ and $t \ge 0$. Remember also that $P = 7x + 11y$ must be written in the form $P - 7x - 11y = 0$.

In the tableau, the pivots are shown in bold type. You are advised to *ring* the pivots.

P	x	y	s	t	l	Equation	
1	−7	−11	0	0	0	①	
0	2	4	1	0	60	②	
0	**3**	3	0	1	60	③	3 is the pivot in the x column.
1	0	−4	0	$\frac{7}{3}$	140	④ = ① $+\frac{7}{3} \times$ ③	
0	0	**2**	1	$-\frac{2}{3}$	20	⑤ = ② $-\frac{2}{3} \times$ ③	2 is the pivot in the y column.
0	1	1	0	$\frac{1}{3}$	20	⑥ = $\frac{1}{3} \times$ ③	
1	0	0	2	1	180	⑦ = ④ $+ 2 \times$ ⑤	
0	0	1	$\frac{1}{2}$	$-\frac{1}{3}$	10	⑧ = $\frac{1}{2} \times$ ⑤	
0	1	0	$-\frac{1}{2}$	$\frac{2}{3}$	10	⑨ = ⑥ $-\frac{1}{2} \times$ ⑤	

The maximum value is 180, when $x = 10$ and $y = 10$.

Here is a commentary on the solution to Example 8.3.1. For Steps 1 and 2, the original equations are written as the first three rows of the tableau.

Step 3   The x-column is chosen as one which has a negative top element.

Step 4   In the x-column, $a_0 = -7$ so it can't be the pivot. Looking at the others,

$a_1 = 2$, $l_1 = 60$ so $\dfrac{l_1}{a_1} = 30$; $a_2 = 3$, $l_2 = 60$ so $\dfrac{l_2}{a_2} = 20$. As 20 is the least,

$a_2 = 3$ is the pivot.

Step 5   Equation 3 is then divided by the pivot, and labelled Equation 6. (It is put in this position so that the essential order of Equations 1, 2 and 3 is maintained.)

Step 6   Then Equation 4 and Equation 5 are derived from Equation 1 and Equation 2 with their x-coefficients made 0 by subtracting appropriate multiples of Equation 3.

When you have derived Equations 4, 5 and 6, you have completed Step 6 and carried out one iteration of the Simplex algorithm. It is conventional to rule a line under the equations at this stage.

The top equation is now $P - 4y + \frac{7}{3}t = 140$, where $y \geqslant 0$ and $t \geqslant 0$, so you can make P larger by increasing y.

Step 7   At this stage not all the elements in the top row (Row 4) are non-negative: there is an element -4. So you have to return to Step 3 and choose another column.

Step 3   The y-column is chosen, as the number at the top, -4, is negative.

Step 4   This time 2 is the pivot, because $\frac{20}{2} = 10$ is less than $\frac{20}{1} = 20$, and -4 is not a candidate for the pivot because it is negative.

Step 5   Equation 5 is then divided by the pivot, and becomes Equation 8, in the same position as Equation 5.

Step 6   Then Equation 7 and Equation 9 are derived from Equation 4 and Equation 6, with their y-coefficients made 0 by subtracting appropriate multiples of Equation 5.

Step 7   All the top elements (that is, those in Equation 7) are now non-negative, so you move to Step 8.

Step 8   At this stage, Equation 7, which is $P + 2s + t = 180$, tells you that the maximum value of the objective function is 180. This occurs when s and t both take the value 0. Equation 8, which is $y + \frac{1}{2}s - \frac{1}{3}t = 10$, tells you that this happens when $y = 10$. Equation 9, which is $x - \frac{1}{2}s + \frac{2}{3}t = 10$, tells you that $x = 10$.

The Simplex algorithm can be used to minimise a function with a simple trick: to minimise a linear function, you maximise its negative. Thus to minimise $4x - 5y - 3z$, you maximise $-4x + 5y + 3z$. The minimum for the original problem is then the negative of the maximum.

Example 8.3.2

Minimise $4x - 5y - 3z$ subject to $x - y + z \geqslant -2$, $x + y + 2z \leqslant 3$ and $x, y, z \geqslant 0$.

Start by introducing the slack variables s and t by writing $x - y + z - s = -2$ and $x + y + 2z + t = 3$, where $s \geqslant 0$ and $t \geqslant 0$. Recall that in the tableau the elements in the last column (except possibly the top one) must not be negative. Therefore the first equation is written in the form $-x + y - z + s = 2$.

Next, let $P = -4x + 5y + 3z$, since you are going to minimise $4x - 5y - 3z$ by maximising its negative. The equation $P = -4x + 5y + 3z$ must be written in the form $P + 4x - 5y - 3z = 0$.

Here is the tableau.

P	x	y	z	s	t		Equation
1	4	-5	-3	0	0	0	①
0	-1	**1**	-1	1	0	2	②
0	1	1	2	0	1	3	③
1	-1	0	-8	5	0	10	④ = ① + 5× ②
0	-1	1	-1	1	0	2	⑤ = ②
0	2	0	**3**	-1	1	1	⑥ = ③ − ②
1	$\frac{13}{3}$	0	0	$\frac{7}{3}$	$\frac{8}{3}$	$\frac{38}{3}$	⑦ = ④ + $\frac{8}{3}$ × ⑥
0	$-\frac{1}{3}$	1	0	$\frac{2}{3}$	$\frac{1}{3}$	$\frac{7}{3}$	⑧ = ⑤ + $\frac{1}{3}$ × ⑥
0	$\frac{2}{3}$	0	1	$-\frac{1}{3}$	$\frac{1}{3}$	$\frac{1}{3}$	⑨ = $\frac{1}{3}$ × ⑥

The maximum value of $-4x + 5y + 3z$ is $\frac{38}{3}$, or $12\frac{2}{3}$, so the minimum value of $-4x + 5y + 3z$ is $-12\frac{2}{3}$. This occurs when $x = 0$, $y = \frac{7}{3}$ and $z = \frac{1}{3}$.

The reasons for the values of x, y and z come from examining Equations 7, 8 and 9 in conjunction with the inequalities $x, y, z, s, t \geqslant 0$. Equation 7 is $P + \frac{13}{3}x + \frac{7}{3}s + \frac{8}{3}t = \frac{38}{3}$. This shows that the maximum value of P occurs when $x = s = t = 0$. Equation 8 is $-\frac{1}{3}x + y + \frac{2}{3}s + \frac{1}{3}t = \frac{7}{3}$, and since $x = s = t = 0$ this gives $y = \frac{7}{3}$. Similarly $z = \frac{1}{3}$.

Note that using fractions in the Simplex tableau avoids the danger of rounding errors. Although when a computer is programmed for the Simplex algorithm it works in decimals, to a large number of places to retain accuracy, you should use fractions.

In theory, it is possible for the Simplex algorithm to get caught in an infinite loop. However, in practice, such a difficulty is extremely rare.

But there is a problem which can arise at the start of the method. As you have seen, the idea of the Simplex method is to start at the point where all the control variables are zero and then move to 'better' points. However, suppose that one constraint was $x - y \geqslant 1$. Then the point where $x = 0$ and $y = 0$ is not even in the feasible region, and the process cannot start. Methods have been developed to deal with this kind of difficulty, but they are beyond the scope of this book.

Exercise 8A

1 (a) Maximise $2x + 3y$ subject to the constraints $x + 2y \leqslant 6$, $x + y \leqslant 5$ and $x, y \geqslant 0$ by using the Simplex method. Give the whole tableau in your answer.

    (b) Repeat part (a) with the objective function $2x + y$.

2 Use the Simplex algorithm to maximise $x + 4y$ subject to $x + 3y \leqslant 15$, $2x + y \leqslant 12$ and $x, y \geqslant 0$. Show the whole tableau in your answer.

3 Maximise $5x + 3y - 8z$ subject to the constraints $x - z \leqslant 1$, $x + y \leqslant 2$, $3x + 2y - 4z \leqslant 6$ and $x, y, z \geqslant 0$. Show the whole Simplex tableau.

4 Three processes, I, II and III, are involved in the manufacture of three products, A, B and C. For each product, the manufacturing time (hours) and profit per item (£) are as shown.

Product	I	II	III	Profit
A	1	2	3	120
B	5	1	2	70
C	4	4	1	100

The total available manufacturing times on processes I, II and III are 90, 35 and 60 hours respectively.

    (a) What mix of products yields the greatest profit?

    (b) What assumptions have you made in answering part (a)?

5 The marketing director of an insurance company has a budget of £100,000 for half-page advertisements in three specialist magazines. Details of the circulations and advertising costs of these magazines are as shown.

Magazine	Circulation	Cost (£)
Smart Savings	80 000	1000
Capital Investor	120 000	2000
Money Monthly	200 000	2500

The director is instructed that the average circulation for each advertisement should be at least 100 000 and that no more than £40,000 should be spent on any one magazine. What is the maximum number of advertisements she can place and how can she achieve this?

6 Minimise $x + y - 2z$ subject to $2x + y \geqslant z$, $2x + 5 \geqslant 3z$, $3y + 4z \leqslant 12$ and $x, y, z \geqslant 0$.

8.4 Network problems

You have seen how linear programming is concerned with making something optimal (for example, least costs or greatest profit) subject to various constraints. In earlier chapters you saw a range of network problems which were also about constrained

optimisation. Each of these problems can actually be converted into a linear
programming problem.

Here is a very simple example of the shortest path problem.

Example 8.4.1   Shortest paths

Find the shortest path from node A to
node C in Fig. 8.2.

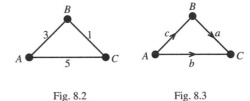

Fig. 8.2   Fig. 8.3

To convert this into a linear
programming question, define a, b
and c by

$$a = \begin{cases} 1 & \text{if } BC \text{ is on the shortest path} \\ 0 & \text{otherwise.} \end{cases}$$

The quantities b and c are defined similarly: see Fig. 8.3.

The objective function is then the total length of the chosen arcs, that is
$a + 5b + 3c$.

Here is the linear programming formulation of the problem.

Minimise $a + 5b + 3c.$

Constraints $b + c = 1,$ There is an arc out of A.
                $a + b = 1,$ There is an arc into C.
                $a = c,$ There is an arc out of B if there is an arc into B.
                $a, b, c \geqslant 0.$

You might well think that each of the final three conditions should take the form $a = 0$
or $a = 1$ rather than $a \geqslant 0$. However, it can be proved that the given conditions are
sufficient and will actually force each variable to be 0 or 1. Before reading on, solve
the linear programming problem and obtain the obvious solution $a = c = 1$, $b = 0$.

Example 8.4.2   Minimum connector

Find the minimum connector for the network of
Fig. 8.4.

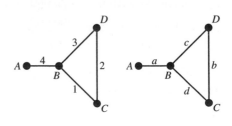

Refer to Fig. 8.5, and define

$$a = \begin{cases} 1 & \text{if } AB \text{ is used} \\ 0 & \text{otherwise.} \end{cases}$$

Fig. 8.4   Fig. 8.5

The quantities b, c and d are defined
similarly.

Here is the problem in a linear programming format.

Minimise $4a + 2b + 3c + d.$

Constraints $a \geqslant 1,$ (A is connected)
              $a + c + d \geqslant 1,$ (B is connected)
              $b + d \geqslant 1,$ (C is connected)
              $b + c \geqslant 1,$ (D is connected)
              $c + d \geqslant 1,$ (A and B are connected)
              and so on, and
              $a, b, c, d \geqslant 0.$

Notice that the constraints omitted above do not add anything new. For example, $a + b + c \geqslant 1$, which arises because B and C are connected, is already covered by $a \geqslant 1$ and $b + c \geqslant 1$.

Again, before reading on, you should solve the linear programming problem. You will obtain the obvious solution $a = b = d = 1, c = 0$.

As you can see, linear programming provides a general problem-solving technique which can be applied to the various network problems studied in this textbook. Nevertheless, the specialised algorithms such as Dijkstra's, Prim's and Kruskal's are usually far easier to apply when dealing with a specific network problem.

Exercise 8B

1 The tableau format for a problem is as shown.

P	x	y	z	r	s	t	
1	−6	−5	−3	0	0	0	0
0	7	7	4	1	0	0	23
0	5	6	2	0	1	0	16
0	4	8	−2	0	0	1	13

(a) Identify the objective function, constraints, control variables and slack variables.

(b) Find the maximum possible value of the objective function and the corresponding values of the control variables.

2 Consider this linear programming problem.

Maximise $3x + 2y$ subject to the constraints $x + y \leqslant 40$, $2x + y \leqslant 50$, $-x + 4y \geqslant 20$ and $x, y \geqslant 0$.

(a) Use a graphical method to solve this problem.

(b) Minimise $3x + 2y$ subject to the same constraints.

(c) What difficulty would arise if you attempted to use the Simplex algorithm to tackle either part (a) or part (b)?

(d) Replace the second constraint by $2x + y \geqslant c$. Plot a graph of the maximum value of $3x + 2y$ against c.

3 A company is producing three models of personal computer for the Christmas market, the Supremo 64, the Supremo 128 and the Supremo 256. Each computer requires a standard microprocessor of which 3000 can be purchased and marketing information has led to the decision that at least five-sixths of the computers should be Supremo 64s. The company can raise £2,000,000 for purchasing components and has 5000 assembly hours available.

Details of assembly times, costs of components and selling prices are as shown in the table.

PC	Assembly time (hrs)	Cost (£)	Selling price (£)
64	1.5	600	1000
128	2	700	1200
256	3	800	1600

Advise the company as to the production strategy which will optimise profits. What assumptions have you made in formulating your advice?

4 (a) Display the following linear programming problem in a Simplex tableau.

Maximise $P = 3x + 3y + 2z$,

subject to $3x + 7y + 2z \leqslant 15$,

$2x + 4y + z \leqslant 8$,

$x \geqslant 0, y \geqslant 0, z \geqslant 0$.

(b) Perform one iteration of the Simplex algorithm, choosing to pivot on an element from the x-column.

(c) State the values of x, y, z and P resulting from the first iteration. How do you know whether or not you have found the optimal solution?

(d) Complete the solution to this problem.

5* Consider the various linear programming problems which you have solved using the Simplex method. For any such problem let n be the number of variables (both control and slack), let m be the number of non-trivial constraints and let v be the number of non-zero variables (both control and slack) in the optimum solution.

Construct a table of values of n, m and v. What do you notice?

Miscellaneous exercise 8

1 Consider the linear programming problem:

maximise $P = 4x + 5y + 3z$,

subject to $8x + 5y + 2z \leqslant 4$,

$x + 2y + 3z \leqslant 1$.

(a) Set up a Simplex tableau for this problem.

(b) Choose as first pivot an element in the x-column. Identify the values of the variables and the objective function after the first iteration.

(c) Complete the Simplex method, explaining how you know that your solution is optimal.

2 The problem in Miscellaneous exercise 7 Question 5 can be formulated as
maximize $P = 4000x + 6000y$ subject to $x + y \leqslant 5000$, $-x + 4y \leqslant 0$, $x + 3y \leqslant 6000$,
$x \geqslant 0$ and $y \geqslant 0$.

(a) By introducing slack variables, set up the problem as an initial Simplex tableau.

(b) Perform one iteration of the tableau, beginning by choosing to pivot on the x-column. Interpret the result of this iteration.

(c) Explain how you know, from the tableau, that the optimal solution has not yet been reached. (OCR)

3 A company manufactures three kinds of robot, the Brainy, the Superbrainy and the Superbrainy X. After production, each robot is tested for its coordination and its logic. The company wants to maximize its profit from the sale of the robots. You may assume that every robot manufactured is sold. The table below shows the times required by these tests, the time available each week, and the profit per robot.

	Coordination test (hours)	Logic test (hours)	Profit per robot (£)
Brainy	6	3	2400
Superbrainy	3	4	3000
Superbrainy X	1	5	3200
Time available (hours)	60	60	

(a) Set up this problem as a linear programming problem.

(b) Convert this problem into an initial Simplex tableau.

(c) Perform one iteration of the tableau, explaining your method carefully. Interpret your solution in terms of the original problem. (OCR, adapted)

4 Phil decides to use an interlocking shelving system. The shelving system can be made up from any combination of narrow shelves, medium shelves and wide shelves.

Suppose that Phil buys x metres of narrow shelves, y metres of medium shelves and z metres of wide shelves.

The problem of maximizing the storage capacity, when using the shelving system, gives the LP formulation.

Maximize $P = 3x + 4y + 5z$,
subject to $2x + 8y + 5z \leqslant 3$,
$\qquad\qquad 9x + 3y + 6z \leqslant 2$,
and $\qquad x \geqslant 0, y \geqslant 0, z \geqslant 0$.

(a) Set up an initial Simplex tableau for this problem. Perform two iterations, choosing to pivot first on an element chosen from the z column.

(b) Interpret the result of each of the two iterations carried out in part (a).

(c) Explain how you know whether or not the optimal solution has been achieved. (OCR)

5 Consider the linear programming problem.

Maximise $P = 2x + y$,

subject to $x + y \leqslant 7$,

$x + 2y \leqslant 10$,

$2x + 3y \leqslant 16$,

$x \geqslant 0$ and $y \geqslant 0$.

(a) By introducing slack variables, represent the problem as an initial Simplex tableau.

(b) Perform one iteration of the Simplex algorithm, choosing to pivot first on an element chosen from the x-column.

(c) State the values of x, y and P resulting from the iteration in part (b).

(d) Explain how you know whether or not the optimal solution has been achieved. (OCR)

6* Consider the following linear programming problem.

Maximize $P = 3x + y$,

subject to $x + 3y \geqslant 10$,

$y + 5 \leqslant 2x$,

$3x + 2y \leqslant 20$,

and $x \geqslant 0$, $y \geqslant 0$.

(a) Graph the constraints, using $0 \leqslant x \leqslant 10$ and $0 \leqslant y \leqslant 10$. Identify the feasible region by shading those regions where the constraints are not satisfied.

(b) Draw and clearly label the line $P = 25$ on your graph.

(c) Use your graph to find the values of x and y that solve the linear programming problem, giving your answers correct to 1 decimal place.

The linear programming problem can be written in the form below (you are not expected to demonstrate this).

Maximize $R = 8X + 6Y + 900$,

subject to $-7X + Y + S = 0$,

$7X + 9Y + T = 350$,

and $X \geqslant 0, Y \geqslant 0, S \geqslant 0, T \geqslant 0$,

where $R = 70P$, $X = 7(3x - y) - 60$, $Y = 7(x + 3y) - 70$, and S and T are slack variables.

(d) Use the Simplex algorithm to calculate the optimum value of R, and the values of X and Y for which it is attained, choosing to pivot first on an element in the X column. Show sufficient working to make your method clear.

(e) Use your answers from part (d) to calculate the optimum value of P, and the values of x and y for which it is attained for the linear programming problem given at the start of the question. (OCR)

Revision exercise

1

	A	B	C	D	E	F	G
A	–	10	13	7	9	–	–
B	10	–	11	–	–	11	–
C	13	11	–	5	–	8	16
D	7	–	5	–	4	–	21
E	9	–	–	4	–	–	21
F	–	11	8	–	–	–	7
G	–	–	16	21	21	7	–

The matrix represents the distance in km along direct routes between the towns A to G.

(a) Copy and complete the network, shown in the figure below, corresponding to the direct route network.

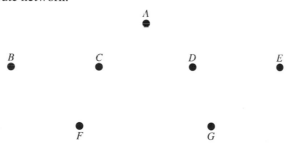

(b) Use Dijkstra's algorithm to identify the route with the least distance between town A and town G. Ensure that you indicate clearly the order in which the vertices are assigned permanent labels and state the minimum distance. (OCR)

2 The network of roads in Question 1 have to be inspected regularly to identify when the road surface needs repairing. Use an appropriate algorithm to determine the least distance which needs to be travelled in the process of inspecting the road surfaces. (OCR)

3 A travelling salesperson lives at the town labelled A in Question 1. The salesperson has to visit each town and wishes to do this in the minimum distance.

(a) Find an upper bound for the distance the salesperson needs to travel, and identify a route with this total distance.

(b) Explain how you might find a lower bound for the minimum distance. (You are not expected to find this lower bound.) (OCR)

4 The outcome of modelling an optimisation problem leads to the following linear programme.

        Cost constraint $180u + 100v + 250w + 200x \leqslant 2500$

        Weight constraint $5u + 5.8v + 6.2w + 2.5x \leqslant 100$

        Objective function $P = 3u + 2v + 4w + 2x$

(a) The modeller decides to use the Simplex method to maximize P. Write down the initial tableau.

(b) Perform one iteration of the Simplex method using column w as the pivot column to determine the next tableau. (Make the process you adopt clear.)

(c) Explain how you know whether or not your tableau is optimal.

(d) The modeller uses a computer to solve the problem and gives the following solution:

    $P = 45.0368;\ u = 8.2721,\ v = 10.1103;\ w = 0;\ x = 0$

and the slack variables are zero.

(i) Interpret the statement about slack variables.

(ii) The variables u, v, w, x represent the number of Igloo, Winter, Camp and Mountain tents which a group preparing for the Duke of Edinburgh Award intend to buy. Explain why the solution presented by the modeller is not relevant.   (OCR)

5 The group mentioned in Question 4 decide to limit themselves to the Winter and Mountain tents. These cost £100 and £200, and weigh 5.8 kg and 2.5 kg respectively. The group has £750 to spend on tents and wants to restrict the total weight of the tents to 25 kg.

(a) They buy v Winter tents, and x Mountain tents. Write down four inequalities satisfied by v and x.

(b) Represent the inequalities you have written in part (a) on graph paper.

(c) Each tent is large enough for two people. There are nine people in the group. Write down another inequality satisfied by v and x. Represent this on your graph paper.

(d) Use your graph to decide how many of each tent the group should buy. State the maximum amount of money the group will have left.   (OCR)

6 An express delivery pizza company promises to deliver pizzas within 30 minutes of an order being telephoned in. Four customers telephone in orders at the same time.

While the pizzas are cooking, which takes 10 minutes, a delivery route must be planned. All four pizzas will be delivered by the same person.

The figure shows the pizza company (P) and the four customers (A, B, C and D). The table shows the travel times for each possible leg of the journey (excluding stopping to deliver the pizza).

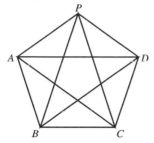

	P	A	B	C	D
P	–	5	3	4	2
A	5	–	3	4	4
B	3	3	–	3	5
C	4	4	3	–	5
D	2	4	5	5	–

(a) Find an upper bound for the travelling salesperson problem.

(b) Delete A from the network, and hence find a lower bound to the travelling salesperson problem.

It takes $2\frac{1}{2}$ minutes to stop and deliver a pizza.

(c) Explain why a travelling salesperson solution that takes 16 minutes does not guarantee that everyone will get their pizza within 30 minutes of their telephone call.

(d) In reality, what happens is that the customer who is nearest (in the sense of having the shortest travel time) gets their pizza first, then the customer who is nearest to the first one, then the customer who is nearest to the second one, and finally the remaining customer.

Write down the order in which the pizzas are delivered using this rule, and work out how long the fourth customer has to wait for their pizza.   (OCR, adapted)

7 Consider the linear programming problem:

     maximize $P = x + 2y$,

     subject to $x + y \leqslant 5$,

                $x + 4y \leqslant 10$,

                $y - x \leqslant 1$,

                $x, y \geqslant 0$.

(a) Represent the constraints graphically, shading the regions where the inequalities are **not** satisfied.

(b) Find the coordinates of the vertices of the feasible region.

(c) Represent the problem as an initial Simplex tableau.

(d) Perform **two** iterations of the Simplex algorithm, choosing to pivot first on an element chosen from the x-column.

(e) State the values of x, y and P resulting from each iteration in part (d).   (OCR)

8 The figure shows a city centre, X, and eight suburbs, A to H. The eight suburbs and the city centre are to be connected using an overhead monorail system. The arcs in the figure represent the possible monorail links.

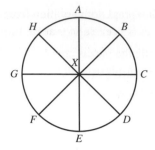

The costs (in £ millions) of the possible links are shown in the table.

	A	B	C	D	E	F	G	H	X
A		2						3	4
B	2		4		–	–		–	5
C		4		3					3
D			3		7				5
E				7		5			6
F					5		4		5
G		–	–	–		4		6	3
H	3					–	6		2
X	4	5	3	5	6	5	3	2	

(a) Use a greedy algorithm, starting from suburb A, to find the minimum connector tree that links the eight suburbs and the city centre at minimum cost.

(b) Work out the total cost of the monorail links in the minimum connector tree found in part (a).

The times (in minutes) that it would take to travel each of the proposed monorail links are shown in the following table.

	A	B	C	D	E	F	G	H	X
A		5				–		8	7
B	5		10						8
C		10		3					6
D			3		6				5
E				6		4			7
F					4		4		5
G				–		4		6	4
H	8						6		7
X	7	8	6	5	7	5	4	7	

(c) Work out the time taken to reach the city centre from each of the suburbs using only the monorail links in the minimum connector tree found in part (a), and ignoring the time for which the train is stationary.

(d) Adapt your solution from part (a) to find a way of linking the eight suburbs and the city centre so that the time taken to reach the city centre from each suburb is no more than 15 minutes, and the total cost is less than £30 million.

9 A planar graph is one that can be drawn so that no two arcs cross each other.

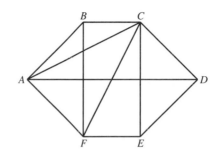

(a) Show that the graph in the figure is planar by redrawing it so that no two arcs cross each other.

An algorithm for deciding whether a graph is planar or not is described below.

Step 1 Find a path that passes through every vertex, starting and ending at the same vertex (a cycle) and which crosses no other arc.
In the example above, you could choose *ABCDEFA*.

Step 2 Choose any arc in the original graph which is not included in the cycle. Construct two lists. Initially the left hand list contains the chosen arc and the right hand list is empty.
In the example above, you could choose *AD*.

Step 3 Add to the right hand list all arcs which cross any arc in the left hand list and which are not included in the cycle.

Step 4 Add to the left hand list all arcs which cross any arc in the right hand list and which is not included in the cycle.

Step 5 Repeat Steps 3 and 4 until no more arcs can be added to either list.

Step 6 If there are any arcs which are not in the cycle but are not yet included in the lists, choose one such arc, add it to the left hand list and go back to Step 3.

If any arc occurs in both lists then the graph is non-planar.

Otherwise, a planar graph can be drawn by putting all the arcs in the left hand list inside the cycle, and all the arcs in the right hand list outside the cycle.

(b) Demonstrate the use of this algorithm, on the graph in the figure above, by starting with the cycle *ABCDEFA* in Step 1, and choosing the arc *AD* in Step 2.

(c) The graph in the figure alongside is non-planar. Show what happens when you apply the algorithm to this graph.

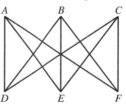

(d) The graph in the figure alongside is planar. Explain why the algorithm cannot be used to show that this graph is planar.   (OCR)

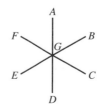

10 The network in the figure represents the rail links between nine cities. The time taken to travel each section of track is shown in minutes.

(a) Use Dijkstra's algorithm, on a copy of the figure, to find the quickest route from R to Z. Show all your working clearly, and indicate the order in which you assign permanent labels to the vertices.

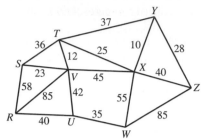

(b) Use a greedy algorithm, starting at R, to construct the minimum connector (minimum spanning tree) for the network in the figure. Show all your working clearly and give the order in which cities are added to the tree.

State the total length in minutes of the minimum connector.

(c) Explain why the length of the minimum connector must be less than the length of the optimal solution to the travelling salesperson problem for this network.   (OCR)

11 Amy, Bea, Cara and Di are four friends. The lengths of the paths between their houses, measured in kilometres, are given in the table.

	A	B	C	D
A	0	3.6	2.2	2.3
B	3.6	0	1.2	1.0
C	2.2	1.2	0	1.9
D	2.3	1.0	1.0	0

Amy wants to visit each of her friends and then return to her own house using the shortest route possible.

She decides to use the following algorithm:

She will start at her own house.

At each stage she moves to a friend's house that she has not yet visited; she always chooses the nearest such house.

She repeats this until all three friends have been visited. She then returns to her own house.

(a) Write down the route that this algorithm gives for Amy, and find the length of this route.

(b) By considering all possible ways of putting the girls into two pairs, find the length of the shortest route that uses all six paths between the four houses.

12 Tommy has three favourite types of breakfast cereal. He enjoys eating each of these cereals, or a mixture of two or more of the cereals. The nutritional content in 100 g of each of the three cereals is shown in the table.

	Arctic Flakes	Banana Bran	Crispy Crunch
Energy	500 kcal	450 kcal	350 kcal
Protein	10 g	10 g	6 g
Fibre	4 g	9 g	7 g
Carbohydrates	48g	60g	68 g
Fats	20 g	2 g	10 g

Let $100a$ g be the amount of Arctic Flakes, $100b$ g be the amount of Banana Bran and $100c$ g be the amount of Crispy Crunch that Tommy eats for breakfast.

Tommy's breakfast must provide him with at least 3.5 g of protein and at least 2 g of fibre.

(a) Write down two constraints that must be satisfied by the variables a, b and c, other than that they must be non-negative.

Tommy's breakfast must provide him with no more than 24 g of carbohydrates and no more than 5 g of fats.

(b) Write down two more constraints that must be satisfied by the variables a, b and c, other than that they must be non-negative.

Tommy is on a diet, so he wants to minimize the energy value of his breakfast.

(c) Write down the objective function for this problem.

(d) Give two reasons why this problem could not be solved immediately in this form using the Simplex algorithm.

Tommy's breakfast should be a 40 g portion of cereal.

(e) Show how this constraint can be incorporated into the problem, and explain how the resulting problem could then be solved, without using the Simplex algorithm.   (OCR)

Mock examination 1

Time 1 hour 20 minutes

Answer all the questions.
Graphic calculators may not be used.

1 £20,000 is to be invested in three different forms of savings X, Y and Z. The expected annual returns from each are 4%, 7% and 11% respectively.

It is required that no more than half the money should be invested in Z, the riskiest investment, and that the expected annual return should be at least 8%.

Suppose that £x and £y denote the amounts invested in X and Y, respectively. Formulate the problem of maximising the amount of money in X, an instant access account, as a linear programming problem in the variables x and y. [5]

2 Use an appropriate algorithm to find a path of least weight from A to B in the network shown. To demonstrate your application of the algorithm, show all temporary labels.

Give the least weight path and its total weight. [6]

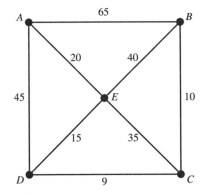

3 (i) For a connected graph with n nodes, state the number of arcs in a minimum spanning tree. [1]

The following matrix shows the distances, in miles, between six towns.

	A	B	C	D	E	F
A	-	11	19	15	16	22
B	11	-	21	8	12	9
C	19	21	-	15	13	14
D	15	8	15	-	21	11
E	16	12	13	21	-	14
F	22	9	14	11	14	-

(ii) Use an algorithm to find the minimum spanning tree and its length. Show your working in full, and draw the spanning tree. State the order in which arcs are chosen. [5]

4 The graph shown in the figure represents a road system. The lengths of the roads are shown in metres.

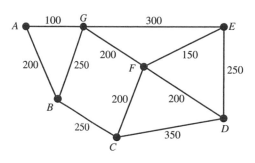

   (i)   Explain why the graph is not Eulerian.   [1]

  (ii)   Find a shortest route, starting and finishing at A, which traverses each road at least once. State the length of your route.   [5]

 (iii)   It is also required to find the shortest route, starting and finishing at A, which visits each node. Find upper and lower bounds for the optimal solution, briefly justifying your answers.   [7]

5 (i)   Use the Shuttle Sort algorithm to put the following numbers into order of increasing size.

$$3, 5, 1, 9, 8, 2.$$

      Show the result on the list of each pass and complete a table of comparisons and swaps required at each pass. Find the total number of comparisons and the total number of swaps.   [6]

  (ii)   For a list with six elements, calculate the maximum possible numbers of comparisons and swaps. Give an example of a list requiring these maximum numbers of comparisons and swaps.   [3]

 (iii)   Determine the maximum numbers of comparisons and swaps needed for a list with n elements. What is the order of the Shuttle Sort algorithm?   [3]

 (iv)   A computer implementation of the Shuttle Sort algorithm takes 2 seconds to sort a list of 100 elements. Estimate the time it would take to sort a list with 1000 elements.   [2]

6 Consider the following linear programming problem.

    Maximise $P = 3x + 2y,$

    subject to $x + y \leqslant 8,$

                 $x + 3y \leqslant 21,$

                 $4x + y \leqslant 30,$

                 $x \geqslant 0, y \geqslant 0.$

   (i)   Graph the constraints, identifying the feasible region by shading the unwanted regions. Draw and clearly label the line $P = 30$.   [4]

  (ii)   Find the values of x and y that solve the linear programming problem, giving your answers in exact form.   [2]

 (iii)   Write down an initial Simplex tableau for this problem. Perform one iteration of the Simplex method, choosing to pivot on an element in the x-column. Identify the point on your graph in part (i) which corresponds to the result of your iteration.   [5]

 (iv)   Complete the Simplex method to find the maximum value of P. Give the values of the slack variables at the maximum point and say what information is given about each inequality by the value of its slack variable.   [5]

Mock examination 2

Time 1 hour 20 minutes

Answer all the questions.
Graphic calculators may not be used.

1 Consider the linear programming problem:

maximise $P = 3x + y,$
subject to $2x + 3y \leqslant 6,$
              $x + 2y \leqslant 4,$
              $x \geqslant 0, y \geqslant 0.$

(i)   Represent this problem as an initial Simplex tableau.   [2]

(ii) Carry out one iteration of the Simplex algorithm, choosing to pivot on an element in the x-column.   [3]

(iii) State the maximum value of P, explaining how you know that the maximum has been achieved.   [2]

2 A small factory has a licence to produce two types of digital recorder, VCRs and DVDs. Both types require the use of two assembly lines A and B. Each VCR requires 60 minutes on line A and 30 minutes on line B. Each DVD requires 60 minutes on each line. Each week, assembly line A is available for 48 hours, and line B is available for 40 hours.

The factory makes a profit of £30 on each VCR and £90 on each DVD. To fulfil a contract, the factory must produce at least 10 VCRs each week. Suppose that the factory makes x VCRs and y DVDs each week. It is required to maximise the weekly profit.

(i)   Set up the problem as a linear programming problem with an objective function and a set of constraints.   [4]

(ii) Represent the constraints graphically, shading the region where the constraints are not satisfied.   [3]

(iii) Calculate the coordinates of the vertices of the feasible region. Hence find values of x and y that will maximise the profit. Calculate this profit.   [4]

(iv) What difficulty would there have been in applying the Simplex method to this problem?   [1]

(v)   Suppose that the contract had required at least 15 VCRs per week. Find the best integer solution in this case.   [2]

3 The diagram represents a network of footpaths. The distances are in hundreds of metres.

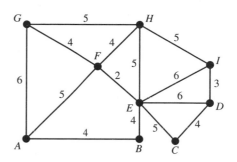

(i) Apply Dijkstra's algorithm to find the shortest path from *A* to *I*. Show all your working clearly, and state the length of the shortest path. **[4]**

(ii) Write down the shortest distances from *A* to *D*, *G* to *D*, and *G* to *I*. **[3]**

To clear the footpaths of litter, a warden must walk along each footpath.

(iii) Explain why some footpaths will have to be repeated. **[2]**

(iv) Apply the Chinese Postman algorithm to find a shortest route that starts and finishes at *A* and travels each footpath at least once. Write down a suitable route and give its total length. **[4]**

4 An algorithm that can be applied to the nodes of a graph is defined as follows.

Step 1 Choose any node and add it into set *V*.

Step 2 Add all nodes adjacent to any just added to *V* into set *W*.

Step 3 Add all nodes adjacent to any just added to *W* into set *V*.

Step 4 If *V* has been altered by Step 3, then return to Step 2. Otherwise stop.

(i) Apply this algorithm to the following graphs, in each case starting with node *a*.

(a) 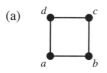 (b) 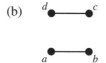 (c)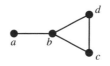

[6]

Show the result of each step in a table:

Step	*V*	*W*
1	*a*	-
2	-	-
⋮	⋮	⋮

(ii) Explain how this algorithm can be used to determine if a graph is connected. **[2]**

(iii) Explain how this algorithm can be used to determine if a connected graph is bipartite. **[2]**

(iv) Describe the result of applying the algorithm to

   (a) a cycle with an even number of nodes,

   (b) a cycle with an odd number of nodes.

   What can you say about any cycles of a bipartite graph? **[3]**

5 The distances, in kilometres, between four towns are as shown.

	A	B	C	D
A	-	110	90	70
B	110	-	80	90
C	90	80	-	60
D	70	90	60	-

(i)   Use Prim's algorithm, starting from A, to find the length of the minimum connector for the four towns. Show all your working clearly and draw the minimum spanning tree. State the order in which arcs are chosen.   [5]

A politician's constituency, H, is at distances 80, 90, 50 and 70 km from A, B, C and D respectively. The politician has to start from H and visit each town before returning to H.

(ii)   Apply the Nearest Neighbour algorithm from H to find a solution to this travelling salesperson problem.   [2]

(iii) Find an improvement upon the solution of part (ii) and explain why your new solution is an upper bound for the optimal solution.   [3]

(iv)   Use your answer to part (i) to find a lower bound for the optimal solution. Explain your answer fully.   [3]

Answers

1 Algorithms

Exercise 1A (page 5)

1 (a) 10 comparisons, 10 swaps
   (b) 10 comparisons, 10 swaps
   (c) For lists in reverse order, both algorithms require the same number of comparisons and swaps.

2 (a) 15 (b) $\frac{1}{2}n(n-1)$

3 4 1 6 8 2, 1 4 6 8 2, 1 4 6 8 2, 1 4 6 8 2, 1 2 4 6 8
   7 comparisons, 4 swaps

4 (a) $n-1$ (b) 10 comparisons, 2 swaps plus any that you make in your own algorithm.

5 (a) 7, the sum of m and n.
   It will work only if m is a positive integer. The value of n does not matter.
   (b) 12, the product of m and n.
   It will work only if m is a positive integer. The value of n does not matter.

Exercise 1B (page 8)

1 (a) 4
   (b) In three lengths: $5+3$, $4+4$, $2+2+2+1$

2 (a) 10^{-3} s (b) 1 s   (c) $16\frac{2}{3}$ minutes
   (d) 3×10^{287} years

3 Lane 1: 14 11
   Lane 2: 9 5 5 4
   Lane 3: 4 4 4 3 3 3 3

4 For example,
   120 cm, 40 cm, 40 cm, 50 cm, 50 cm, 100 cm

5 (a)

Step	1	2	4	1	2	4	1	2	4	1
m	3	5	2	2	3	1	1	2	1	1
n	5	3	3	3	2	2	2	1	1	1

   (b) 6. Make the rectangle into two 3×3 squares with three 2×2 underneath.

Miscellaneous exercise 1 (page 12)

1 (a) 1.414 215 686, 1.732 050 81, 2.236 068 896
   (b) It finds $\sqrt{x}$.

2 (a) (i) There are two roots.
   (ii) There is a repeated root.
   (iii) There are no roots.
   (b) It describes the roots of the equation $ax^2+bx+c=0$.

3 (a) (i) 3 (ii) 2 (iii) 1
   (b) The greatest common factor of X and Y.

4

X	11	5	2	1	0
Y	9	18	36	72	72
T	0	9	27	27	99

   It multiplies X and Y.

5 (a) 8; 0
     8,12; 1
     2, 8, 12; 1
     2, 8, 12, 54; 3
     2, 8, 12, 23, 54; 4
     2, 8, 12, 23, 31, 54; 5
   (b) 2, 8, 12, 23, 31, 54
   (c) $\frac{1}{2}n(n-1)$, quadratic, or order n^2

6 (a) $\times 1000$
   (b) $+19.9$ correct to 3 significant figures

7 (a) 7 (b) 6 (c) 5

8 (a)

	C	S	T	D
Start	0.8660	3.0000	3.4641	0.4641
1st	0.9659	3.1058	3.2154	0.1096
2nd	0.9914	3.1326	3.1597	0.0270
3rd	0.9979	3.1394	3.1461	0.0067

   (b) 3
   (c) Looks to be about 0.25.
   (d) 0.4641×0.25^n

9 (a) (i)

I	J	M	$X=S(M)$?	$X<S(M)$?
1	8	4	No	No
5		6	No	No
7		7	Yes	

   7 is printed.

   (ii)

I	J	M	$X=S(M)$?	$X<S(M)$?
1	8	4	No	No
5		6	No	No
7		7	No	No
8		8	No	Yes
8	7			

   At this stage $I>J$, so FAIL is printed.
   (b) 10, 5, 2, 1
   (c) (i) 4 (ii) 5

10 (a) 3 4 1 5 2

(b) $n-1$

(c) No random number generated is wasted in this algorithm, but many are wasted in the earlier algorithm.

2 Graphs and networks

Exercise 2A (page 21)

1 (a) Eulerian　　　(b) neither

(c) semi-Eulerian　　(d) Eulerian

2 Consider the closed trail containing every arc precisely once. Each time a node occurs it has an arc going in and an arc coming out. (This applies even to the initial and the final nodes if you consider them together.) Each node is therefore at the end of an even number of arcs, and therefore has even order.

3 (a) 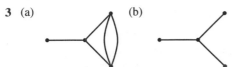 (b)

4 Yes. Start or finish in the top left room or the bottom left room.

5 Two towns linked by a motorway and by one or more local roads.

6 (a) $n-1$　　(b) For odd values of n.

7 (a) rs

(b) $K_{r,s}$ is Eulerian if and only if r and s are both even. It is semi-Eulerian either if $r=1$ and $s=1$, or if $r=2$ and s is odd, or if $s=2$ and r is odd.

8 (a)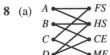
A　　FS
B　　HS
C　　CE
D　　ME

(b) 2, Brian/David → Home Sec./Chancellor

Exercise 2B (page 25)

1

3 There is 1 tree with 3 nodes, 2 trees with 4 nodes, 3 trees with 5 nodes, and 6 trees with 6 nodes.

4 (a) 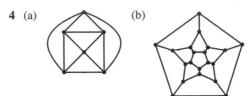 (b)

5

6

n nodes

7 (a) Counting 3 arcs for each of R regions double counts each arc, so $3R=2A$.

(c)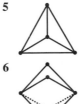

Exercise 2C (page 28)

1 (a) Times from city A to city B, and from city B to city A could depend on road works or congestion, which may affect the two directions differently.

(b) (i) DC circuits　　(ii) AC circuits

2 (a)

$$
\begin{array}{c}
 & A & B & C & D & E & F \\
A & - & - & 1 & 6 & 7 & 8 \\
B & - & - & 2 & 3 & 4 & 3 \\
C & 1 & 2 & - & 7 & - & 2 \\
D & 6 & 3 & 7 & - & 5 & - \\
E & 7 & 4 & - & 5 & - & 9 \\
F & 8 & 3 & 2 & - & 9 & -
\end{array}
$$

(b)

				To			
		A	B	C	D	E	F
	A	−	−	1	−	7	−
	B	−	−	2	3	4	3
From	C	−	2	−	7	−	2
	D	6	3	−	−	5	−
	E	7	4	−	5	−	9
	F	8	−	−	−	9	−

3 (a) Use the two arcs of weight 2, the three arcs of weight 3 and the arc CF of weight 4, giving a total of 17.

(b) The route is $ACDEG$ making a total of 10.

4 $H=2C+2$

5

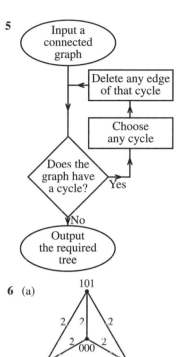

6 (a)

```
        101
         △
    2  ? \  2
   2/ 000  \2
  110    2   011
```

(b) It is possible to tell if a single error has occurred in the transmission.

7 (a) $\sum h_i = 2H$

(b) The sum of all the orders is $2A$.

Miscellaneous exercise 2 (page 29)

1 (a)

	Faces	Vertices	Edges
Cube	6	8	12
Pyramid	5	5	8
Tetrahedron	4	4	6

(b) $F + V = E + 2$. Each shape can be made into a planar graph for which Euler's relation applies.

2 (a) There are two odd nodes, the two islands on the right, so the trail must start on one of these islands and finish on the other.

(b) No

3 (a) *ABCDEA* is a cycle. The chords crossing the circle are *AD*, *BD*, *BE* and *CE*. Call these *x*, *y*, *z* and *t*. There are now arcs joining *x* to *z* and *t*, and *y* to *z* and *t*. This is a bipartite graph, with *x* and *y* in one set, and *z* and *t* in the other. Choosing *x* and *y*, that is, *AD* and *BD*, you can draw them outside the 'pentagon' to make a planar graph.

(b) Choose *ABCDEFA* as the cycle. The chords crossing the circle are *AD*, *BE* and *CF*. Call these *x*, *y* and *z*. The new graph is non-bipartite, so the original graph is not planar.

(c) There is no cycle which includes *C*, so stop. (However, the original graph is clearly planar.)

3 Minimum connector problems

Exercise 3A (page 35)

1 You should find 16 spanning trees.

2 (a) There are 14 spanning trees.

(b) The minimum spanning tree contains the arcs with weights 27, 28, 31, 36, 43. The weight is 165.

(c) The order of choice is 43, 27, 31, 28, 36.

3 The cheapest way costs £1500, and consists of the arcs *HL*, *ML*, *LC*, and *CI*.

5 The arcs used are two of weight 3, one each of weight 5 and 6, two of weight 7 and one of weight 8. The total length is 39.

6 (a) When its removal would disconnect the graph.

(b) Step 1   Find the arc of greatest weight whose removal does not disconnect the graph. Stop if no such arc exists.

Step 2   Remove this arc from the graph and return to Step 1.

Exercise 3B (page 39)

1 (a) AB, BH, BG, AC, CE, CD, AF in that order.
       Total length 215 km.

   (b)

	A	B	C	D	E	F	G	H
A	–	20	27	50	60	30	24	50
B	20	–	50	60	50	65	30	30
C	27	50	–	30	10	80	80	75
D	50	60	30	–	33	70	85	100
E	60	50	10	33	–	85	75	55
F	30	65	80	70	85	–	70	100
G	24	30	80	85	75	70	–	45
H	50	30	75	100	55	100	45	–

   The minimum connector consists of the arcs
   AB, AG, AC, CE, BH, AF, CD and takes
   171 minutes.

   (c) The two minimum connectors are different.
        The route of least length connecting G to
        other points is 30 km from G to B which
        takes 30 minutes. The connection of shortest
        time is the motorway route of 40 km from G
        to A which takes 24 minutes.

2 (a) Starting from A, the minimum spanning tree
        consists of the following arcs, added in this
        order: AB, BD, DE, DC, EF. The total cost is
        £240.

   (b) This has no effect because the total cost of
        each spanning tree is increased by the same
        proportion.

   (c) It is now better to connect C to A, rather
        than to D. The total cost is £300.

3 (a) The minimum connector is, in order: Paris –
        Orléans – Tours – Le Mans – Poitiers –
        Dijon – St-Etienne – Grenoble – Marseilles
        – Nice. The length is 1574 km.

   (b) The length is increased by only 99 km.
        Geneva is linked to Grenoble and Dijon, and
        the Dijon – St-Etienne link is dropped.

Miscellaneous exercise 3 (page 45)

1 New York – Washington – Chicago – Dallas –
   Denver – Los Angeles; 3150 miles

2 (a) The arcs are: BF, BD, BA, AC, CG, CE. The
        length is 1420 metres.

   (b) AC is replaced by DC, and the length is
        increased by 100 metres.

3 In order, BF, BD and AC in either order, AB, CG,
   CE.

4 (a) 8 seconds   (b) 200 seconds
   (c) $5\frac{5}{9}$ hours

5 For a connected graph with n nodes,
   Step 1   Choose the arc of greatest weight.
   Step 2   Choose from those arcs remaining the
             arc of greatest weight which does not
             form a cycle with already chosen arcs.
   Step 3   Repeat Step 2 until $n-1$ arcs have been
             chosen.
   Lo – Br, Lo – Bi, N – Li, Bi – Li, Lo – Le,
   Lo – S, Li – Ld, N – M, N – Sh. Total 1273 km.

6 (a) 42 cm   (b) By 7 cm to 35 cm

7 (a) The order is AD, AE, EF, FB, FC, with a
        total cost of £69 .

   (b) The furthest towns on the minimum
        connector are D and C. The route is $DAEFC$
        with a time of 70 minutes, so it is not
        possible, even with B omitted.

   (c) The minimum connector is now, in order
        from A, AE, ED, EF, FB, BC, so the route
        from A to C is $AEFBC$ with a total time of
        50 minutes, and a cost of £57.

8 (a)

	A	B	C	D	E	F	G
A	–	23	20	–	–	–	–
B	23	–	–	12	13	–	–
C	20	–	–	16	–	–	28
D	–	12	16	–	–	11	12
E	–	13	–	–	–	9	–
F	–	–	–	11	9	–	12
G	–	–	28	12	–	12	–

   (b) In order, AC, CD, DF, FE, DB, DG or FG
   (c) 80

4 Finding the shortest path

Questions (page 48)

1 St. Albans – Oxford – Swindon – Bristol, 175 km

2 St. Albans – Slough – Oxford – Cheltenham,
   1 hour 39 minutes

3 St. Albans – Slough – Swindon – Bristol,
   1 hour 56 minutes

Exercise 4 (page 51)

1 $ADFG$, length 13. The labels on G are
   successively 15, 14, 13.

2 $ACDGIJ$, length 19

3 $ACDF$, £40

4

0	1	2	3	6
7	6	3	4	5
8	5	4	5	6
9	12	11	12	7
10	11	10	9	8

5 (a)

```
        1
      ◆
   0◆   ◆2
      ◆    ◆3
        1
```

(b) It finds the minimum number of arcs needed to link node m to node n.

(c) Moore's algorithm is a special case of Dijkstra's algorithm when all arcs have weight 1.

Miscellaneous exercise 4 (page 53)

1 (a) *ACEF*, length 9

(b) *ACDF*, length 8. *D* is permanently labelled before *ACD* is considered.

2 (a) *AEOD*, 110 minutes

(b) Add 10 to the numbers on each arc at *O*. *ABCD*, 120 minutes

3 (a) By applying the algorithm from *N*.

(b) *ADFKN*, length 8; *BEGIKN*, length 8, *CEGIKN*, length 7. *C* is nearest.

4 (a) *AD* costs £35; *AE* costs £30; *ADC* costs £55; *AEB* costs £85

(b) *AB* is best, cost £90; other routes unaltered.

5 (a) 8 tons

(b) Replace 'D + the weight of the arc joining X to Y' by 'the greater of D and the weight of the arc joining X to Y'.

6 *ABDEHI*, length 19

7 $1+2+1+2+2+1+2+1+2+2=16$

8 10 times

9 (b) In order, *U*, 3; *S*, 4; *T*, 5; then *C*, 6 and *L*, 6 in either order.

(c) Connect *HU*, *US*, *UT*, *TL*, *SC*; cost £9000.

(d) It is a minimum connector, and either Prim's or Kruskal's algorithm would have done.

10 (a) *ACDEG*. Order of labelling, *C*, *D or B* in any order, then *F*, *E* and *G*

(b) *A(B)D(F)G*   (c) 3 minutes

11 (a) *ADCEFB*, 160 metres. Order of labelling, *A, D, C, E, F, G, B*

(b) Route *ADCEFB* + *BG* + *GECDA*. Length = 335 metres.

12 (a) (i) Exeter, Bodmin, Truro, Penzance; 117 miles. Labels in order, E, Pl, Bo, T, Pe

(ii) Exeter, Yeovil, Andover, Newbury, Winchester; 153 miles. Labels in order, E, T, Y, Ba, A, N, W

(b) 5 hours 15 minutes, to the nearest minute

13 (a) *ABDG*, 47 miles. Order of labelling, *A, C, B, D, E, F, G*

(b) *ABEF*, 45 miles

5 Route inspection

Exercise 5A (page 61)

1 (a) A closed trail containing every road enters each intersection the same number of times as it leaves the intersection and therefore contributes an even number to the order of each intersection. Each intersection of odd order must therefore be made even by duplicating some roads.

(b) Total of original roads is 1690 metres. $HI + GC = 340$ m; $IG + HC = 440$ m; $IC + HG = 320$ m which is least, so shortest distance is $(1690 + 320)$ m = 2010 m.

(c) All twice, except for *E* and either *B* or *F* which are passed through three times.

(d) 2010 m; the original closed trail passes through all the intersections and is therefore the shortest route for any starting point.

2 (a) (i)

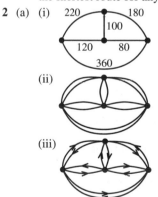

(b) 1060 m

(c) (i) 1360 m   (ii) 2120 m
(iii) 2120 m

3 Represent rooms by nodes and doorways as arcs. Four arcs need to be repeated.

4 (a) The row is 4, $4n-8$, $(n-2)^2$. $2n-4$ need to be repeated.

(b) $2n-2$

Exercise 5B (page 63)

1 (a) Cannot be done as it has two odd nodes.
   (b) Can be done as all the nodes are even.

2 (a) 25 km is the sum of the lengths of all the roads.
   (b) The pairings AC, FG give 5 km;
     AF, CG give 11 km;
     AG, CF give 9 km.
     5 km is lowest, so distance is
     $(25+5)$ km = 30 km.

3 (a) HBC, length 23; HD, length 16;
     HFE, length 24
   (b)

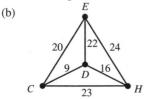

   (c) EF, FH, CD; additional weight $24+9=33$

4 (a) There are seven ways of pairing the first node. For each of these seven ways, there are 15 ways of pairing the remaining six nodes.
   (b) For n odd vertices, $1\times3\times\ldots\times(n-1)$

5 (a) 3770 m   (b) 4520 m

6 Ld – Sh, N – Le, Bi – Br

7 EF and ABC must be repeated.
One path is $ABCEFEDFGBDCBA$.

Miscellaneous exercise 5 (page 65)

1 (a) All the nodes have odd order. To start and finish at A, they must all be of even order, so some arcs must be repeated. Two will need to be repeated.
   (b) Best pairing is $AC+BD=27$ km.
     Total distance is $(86+27)$ km = 113 km.
     One route is $ABDACBDCA$.
   (c) None. All the nodes have order 4 so no pairings are needed.

2 (a) Each arc joins two nodes, so the number of arc ends is even. Suppose that there are r even nodes, and s odd nodes, and that s is odd. Then the total number of arc ends is odd, a contradiction! So the number of odd nodes is even.
   (b) The best pairing of odd nodes is $CE+FI$ totalling 1000 metres. The total inspection route is therefore 6900 metres.

3 (a) Repeat arcs AX, BZ and CY; length of route is 86 km.
   (b) Adding AX, BZ and CY takes 2 additions but there are 6 of these. There are 8 additions to find the total length of the arcs, and finally these two need to be added to find the total length of the route.
   (c) $AV+BW+CX+DY+EZ$ gives 4 additions, and there are $5\times4\times3\times2\times1=120$ of these. 25 arcs give 24 additions, plus 1 to find the total length of the route.
   (d) There are $n!$ sets, where $n!=1\times2\times\ldots\times n$, of $n-1$ additions, plus n^2-1 additions for the n^2 arcs, plus 1 to find the total length of the route, giving $n!(n-1)+n^2$ additions.
   (e) As it involves $n!$, it increases faster than a polynomial function.

4 (a) (i) $BG\ SW$, $BS\ GW$, $BW\ GS$
      (ii) 93 miles
   (b) 15   (c) $1\times3\times5\times\ldots\times(n-1)$

5 (a) 2800 metres   (b) 4600 metres
   (c) No

6 (a) 180 metres, $ABEBCDEFA$
   (b) (i) $ABEBCRSRYSDEFQPQXPA$
      $(200+10\pi)$ metres
      (ii) B or E

7 (a) There are odd nodes.
   (b) $BD\ FG$, $BF\ DG$, $BG\ DF$
   (c) 5.5, 2.5, 5.5
   (d) $ABFBCEFGEGDEDCA$, 22 miles

8 (a) (i) CD, DE, EF
      (ii) $ABCDEDCEFEBFA$
      (iii) $ABEBCDECBFEFA$
   (b) By starting at C or F

6 The travelling salesperson problem

Exercise 6 (page 73)

1 (a) $ABCDA$, $ABDCA$, $ACBDA$
   (b) $ADCBDA$

2 Calais – Orléans – Poitiers – Bordeaux – Toulouse – Marseille – St-Etienne – Lyons – Dijon – Calais; total 2632 km

3 Pembroke Bay – St Sampson Harbour – St Peter Port – St Martin – Airport – Pleinmont Tower – Perelle Bay – Saline Bay – Soumarez Park – Pembroke Bay; total 22 miles

4 (a) £360
  (b) From C or D, the total is £350.
  (c) B

Miscellaneous exercise 6 (page 77)

1 (a) In each case the lower bound is 30.
  (b) 34
  (c) 34 is only possible if the arcs have weights
      4, 5, 8, 8, 9. But the arcs of weight 4 and 5
      form a cycle with one of the arcs of weight
      8.
  (d) $4+5+9+8+9 = 35$

2 The units are kilometres.
  (a) $80+110+230+90+90+110+140 = 850$
  (b) $170+110+135+90+90+110+140 = 845$
  (c) $(170+110)$
$$+(80+90+90+100+110) = 750$$
$$750 \leqslant \text{optimum} \leqslant 845$$

3 (a) $\begin{pmatrix} 40 + 72 + 54 + 67 \\ +61 + 157 + 138 \\ +122 + 123 + 150 \end{pmatrix}$ miles $= 984$ miles

  (b) $\begin{pmatrix} 40 + 54 + 61 \\ +61 + 68 + 72 \\ +138 + 150 \end{pmatrix}$ miles $= 644$ miles

  (c) The optimum solution has length between
      $(122 + 123 + 644)$ miles $= 889$ miles and
      984 miles.

4 (a) 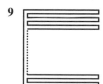 (b) (c)

5 (a) $MEABCDM$ giving a cost of £81;
      $AEMCD$ – not possible;
      $EABMCDE$ giving a cost of £77
  (b) $£(12+13) + £(11+12+13+14) = £75$
      $£75 \leqslant \text{cost} \leqslant £77$
  (c) No effect on the path. The cost of each
      possible path will be increased by £125.

6 $AEMCD(\text{via } M \text{ or } C)BA$,
  $£(11+12+13+14+30+12) = £92$

7 (a) $ABCDEA$, 200 minutes
  (b) Smallest pair of in and out arcs is
      $B \to A \to E$ taking 60 minutes.
      Minimum connector for B, C, D, E is
      20 + 20 + 30 minutes.
      Total is 130 minutes.
  (c) $EBCDAE$, 150 minutes

8

9

10 Colour the nodes in a chequerboard pattern of
   black and white. Then white nodes are directly
   connected only to black nodes and vice versa.
   Any closed trail must therefore have an even
   number of nodes. A Hamiltonian cycle would
   have an odd number n^2 of nodes, so it is not
   possible.

11 (a) Starting from H, Prim's algorithm chooses,
       in order, HG, HI, IF, GE, GD, DC, CA, CB.
       The length of this route is 720 metres.
   (b) The worst situation occurs when you have to
       follow the minimum connector out and back
       to visit all nodes. This, an upper bound, is
       twice the length of the minimum connector,
       namely 1440 metres.
   (c) A better route is $HGDCABEFIH$, of length
       880 metres.

12 (a) The minimum connector is AB, BH, HD,
       DC, DE, EF, FG; time: 20 seconds.
   (b) The time to sprint to the fireworks and back,
       and light all the fireworks is 36 seconds. The
       route between the first and last firework
       takes at least T seconds, so $\theta \geqslant 36 + T$.
   (c) The route between the first and last firework
       takes at most $2T$ seconds, so $\theta \leqslant 36 + 2T$.
   (d) The route $ABHDCEFG$ takes 21 seconds.
       She has 7 more fireworks to light, taking 14
       seconds and a sprint lasting 10 seconds, total
       45 seconds.

7 Linear programming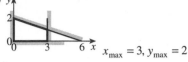

Exercise 7A (page 86)

1 (a)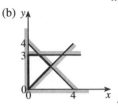
$x_{max} = 3, y_{max} = 2$

(b)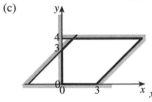
$x_{max} = 2, y_{max} = 3$

(c)
$x_{max} = 7, y_{max} = 4$

(d)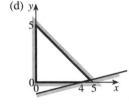
$x_{max} = 4.8, y_{max} = 5$

2 (a) 5 at $(3,1)$ (b) 7 at $(1,3)$
   (c) 15 at $(7,4)$ (d) 10 at $(0,5)$

3 5 at $(2,7)$

4 -4 at $(5,1)$

5 9 at anywhere on the line segment from $(2,3)$ to $(0,9)$

Exercise 7B (page 91)

1 (a) Maximise $7x+8y$, subject to the constraints
   $3x+4y \leqslant 2400$, $x \leqslant 400$, $y \leqslant 400$,
   $x, y \geqslant 0$ are integers.
   (b) £5200 at $(400,300)$

2 £170 at $(10,40)$

3 (a) $p<2.5$ (b) $p>10$

4 (a) Maximise $P = x+y$, subject to
   $2x+4y \leqslant 160$, $5x+2y \leqslant 200$, $x \geqslant 0$,
   $y \geqslant 0$.
   (b) Make 30 of item 1 and 25 of item 2.
   (c) To maximise the profit, make 40 of item 1.

5 (a) $15\frac{5}{17}$ at $\left(2\frac{6}{17}, 3\frac{9}{17}\right)$ (b) 15 at $(0,5)$

6 (a) Maximise $2x+1.7y$, subject to the
   constraints $y \leqslant 1.52x$, $y \geqslant 0.8x$, $x \leqslant 5000$,
   $y \leqslant 3000$, $x \geqslant 0, y \geqslant 0$.
   (b) Make 3750 kg of X, and 3000 kg of Y for a
   profit of £12,600.

7 (a) Maximise x, subject to the constraints
   $5y \geqslant 7x$, $x+y \leqslant 10\,000$, $x, y \geqslant 0$.
   (b) Invest £4167 in a savings account and £5833
   in shares.

8 5 advertisements for the newspapers and 8 for the
   radio. The cost is £10,000 and the audience is
   344 000.

9 Maximise
   $P = 1.25x+0.9y+0.68z+2.25u+1.9v+1.68w$
   subject to $x+u \leqslant 5000$, $y+v \leqslant 10\,000$,
   $z+w \leqslant 10\,000$, $0.7x \leqslant 0.3(y+z)$,
   $0.6y \geqslant 0.4(x+z)$, $0.6v \leqslant 0.4(u+w)$,
   $0.7w \geqslant 0.3(u+v)$.

Miscellaneous exercise 7 (page 93)

1 (a) $44\frac{4}{9}$ ml and $66\frac{2}{3}$ ml
   (b) $133\frac{1}{3}$ ml and $233\frac{1}{3}$ ml

2 (a) 14.47 at $(3.71, 2.35)$
   (b) 14 at $(4,2)$ and $(7,0)$

3 (b) That the past is a guide to the future.
   (c) The pensioner should invest £10,000 in
   deposits, £20,000 in gilts and £40,000 in
   equities.

4 (a)
   (b) (i) $4\frac{12}{17}$ (ii) 7

5 (c) $(0,0)$, $(5000,0)$, $(4500,500)$,
   $\left(\frac{24000}{7}, \frac{6000}{7}\right)$; the corresponding values of
   P are £0, £20 million, £21 million,
   £19 million.
   (d) There should be 4500 UK students and
   500 overseas students.

6 (a) $6x+3y \leqslant 60$, $3x+4y \leqslant 60$, $x \geqslant 0$, $y \geqslant 0$
   (b)
   (c) The company should make 4 Brainy and 12
   Superbrainy robots.

7 (a) $100a+500b+200c \leqslant 2000$
$a+4b+3c \leqslant 50$
(b) $2a+9b+5c$
(c) Maximise $P=2a+9b+5c$,
subject to $100a+500b+200c \leqslant 2000$,
$a+4b+3c \leqslant 50$,
$a \geqslant 0, b \geqslant 0, c \geqslant 0$.
(d) The need for integer solutions.
(e) Another constraint, $a+2b+3c \leqslant 20$

8 (a) Let x be the number of hours he spends preparing for his practical test, and y be the number of hours he spends preparing for his theory test.
(b) Maximise $P=5x+10y$,
subject to $x+y \leqslant 9$,
$5x \geqslant 10$,
$x \leqslant 4$,
$y \geqslant x$,
$x \geqslant 0, y \geqslant 0$.

9 (a) Make x bags of X and y bags of Y.
(b) $4x+4y \leqslant 200$, $6x+3y \leqslant 240$,
$3x+6y \leqslant 240$
(c) $20x+25y$
(d) Integer numbers of bags must be sold.
(e)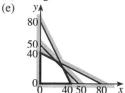
20 bags of X and 30 bags of Y; profit £11.50.
(f) Now $6x+3y \leqslant 240$ has been replaced by $6x+3y=240$, and the $(20,30)$ solution is no longer valid. The solution must lie on the line $6x+3y=240$. The best solution now is 30 bags of X and 20 bags of Y.

8 The Simplex algorithm

Exercise 8A (page 103)

1 (a)

1	-2	-3	0	0	0
0	1	2	1	0	6
0	1	1	0	1	5
1	0	-1	0	2	10
0	0	1	1	-1	1
0	1	1	0	1	5
1	0	0	1	1	11
0	0	1	1	-1	1
0	1	0	-1	2	4

Maximum 11 at $(4,1)$.

(b)

1	-2	-1	0	0	0
0	1	2	1	0	6
0	1	1	0	1	5
1	0	1	0	2	10
0	0	1	1	-1	1
0	1	1	0	1	5

Maximum 10 when $y=0$ and $t=0$. Since $x+y+t=5$, $x=5$, the maximum is 10 at $(5,0)$.

2

1	-1	-4	0	0	0
0	1	3	1	0	15
0	2	1	0	1	12
1	$\frac{1}{3}$	0	$\frac{4}{3}$	0	20
0	$\frac{1}{3}$	1	$\frac{1}{3}$	0	5
0	$\frac{5}{3}$	0	$-\frac{1}{3}$	1	7

Maximum of 20 at $(0,5)$

3

1	-5	-3	8	0	0	0	0
0	1	0	-1	1	0	0	1
0	1	1	0	0	1	0	2
0	3	2	-4	0	0	1	6
1	0	-3	3	5	0	0	5
0	1	0	-1	1	0	0	1
0	0	1	1	-1	1	0	1
0	0	2	-1	-3	0	1	3
1	0	0	6	2	3	0	8
0	1	0	-1	1	0	0	1
0	0	1	1	-1	1	0	1
0	0	0	-3	-1	-2	1	1

The first of the last four rows tells you that the maximum of 8 occurs when $z=0$. The other rows tell you that $x=y=1$. Thus the maximum of 8 occurs at $(1,1,0)$.

4 (a) 10 of A and 15 of B
(b) Two assumptions are that all the items can be sold, and that there are no contracts for C.

5 68; 40 of Smart Savings, 20 of Capital Investor and 8 of Money Monthly

6 Minimum -4 at $(2,0,3)$

Exercise 8B (page 105)

1 (a) $6x+5y+3z$; $7x+7y+4z \leqslant 23$,
$5x+6y+2z \leqslant 16$, $4x+8y-2z \leqslant 13$;
x, y, z; r, s, t
(b) $19\frac{1}{2}$; $x=3$, $y=0$, $z=\frac{1}{2}$

2 (a)   90 at $(10,30)$

(b)   10 at $(0,5)$

(c) $(0,0)$ is not in the feasible region.

(d)

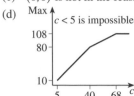

3   Produce 2500 Supremo 64s, 250 Supremo 128s and 250 Supremo 256s. This assumes that all models are sold, and that there are no extra costs and overheads.

4 (a)

1	−3	−3	−2	0	0	0
0	3	7	2	1	0	15
0	2	4	1	0	1	8

(b)

1	0	3	−0.5	0	1.5	12
0	0	1	0.5	1	−1.5	3
0	1	2	0.5	0	0.5	4

(c) $P = 12$, $x = 4$, $y = z = 0$; this is not optimal because there is a negative number in the z-column of the top row.

(d) $P = 15$, $x = 1$, $y = 0$, $z = 6$

5 $v = n - m$. This is usually, but not always, true. It depends on the linear constraints. A proof depends upon knowledge from matrix algebra.

Miscellaneous exercise 8 (page 106)

1 (a)

1	−4	−5	−3	0	0	0
0	8	5	2	1	0	4
0	1	2	3	0	1	1

(b) $P = 2$, $x = \frac{1}{2}$, $y = z = 0$

(c) $P = 2\frac{10}{11}$, $x = \frac{3}{11}$, $y = \frac{4}{11}$, $z = 0$; there are no negative numbers in the top row of the tableau.

2 (a)

1	−4000	−6000	0	0	0	0
0	1	1	1	0	0	5000
0	−1	4	0	1	0	0
0	1	3	0	0	1	6000

(b)

1	0	−2000	4000	0	0	20 million
0	1	1	1	0	0	5000
0	0	5	1	1	0	5000
0	0	2	−1	0	1	1000

This shows that $P = 20$ million when $x = 5000$ and $y = 0$.

(c)   The optimal solution has not been reached because the element at the top of the y-column is negative.

3 (a)   Letting x, y and z be the number of Brainy, Superbrainy and Superbrainy X robots made, the problem is to maximise $P = 2400x + 3000y + 3200z$ subject to the constraints $6x + 3y + z \leqslant 60$, $3x + 4y + 5z \leqslant 60$ and $x, y, z \geqslant 0$ are integers.

(b)

1	−2400	−3000	−3200	0	0	0
0	6	3	1	1	0	60
0	3	4	5	0	1	60

(c)

1	0	−1800	−2800	400	0	24000
0	1	0.5	$\frac{1}{6}$	$\frac{1}{6}$	0	10
0	0	2.5	4.5	−0.5	1	30

The profit is £24,000 at $x = 10$, $y = z = 0$, but the optimum has not yet been reached.

4 (a)

1	−3	−4	−5	0	0	0
0	2	8	5	1	0	3
0	9	3	6	0	1	2

1	−3	−4	−5	0	0	0
0	2	8	5	1	0	3
0	9	3	6	0	1	2
1	$4\frac{1}{2}$	$-1\frac{1}{2}$	0	0	$\frac{5}{6}$	$1\frac{2}{3}$
0	$-5\frac{1}{2}$	$5\frac{1}{2}$	0	1	$-\frac{5}{6}$	$1\frac{1}{3}$
0	$1\frac{1}{2}$	$\frac{1}{2}$	1	0	$\frac{1}{6}$	$\frac{1}{3}$
1	3	0	0	$\frac{3}{11}$	$\frac{20}{33}$	$2\frac{1}{33}$
0	−1	1	0	$\frac{2}{11}$	$-\frac{5}{33}$	$\frac{8}{33}$
0	2	0	1	$-\frac{1}{11}$	$\frac{8}{33}$	$\frac{7}{33}$

(b)   After 1 iteration, $x = y = 0$, $z = \frac{1}{3}$, $P = 1\frac{2}{3}$.

After 2 iterations, $x = 0$, $y = \frac{8}{33}$, $z = \frac{7}{33}$, $P = 2\frac{1}{33}$

(c)   There are no negative entries in the top row, so this is optimal.

5 (a) (b)

1	−2	−1	0	0	0	0
0	1	1	1	0	0	7
0	1	2	0	1	0	10
0	2	3	0	0	1	16
1	0	1	2	0	0	14
0	1	1	1	0	0	7
0	0	1	−1	1	0	3
0	0	1	−2	0	1	2

(c) $x = 7$, $y = 0$, $P = 14$

(d)   The top row in the tableau now has no negative signs, so the optimum has been reached.

6 (a) (b)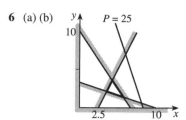

(c) $x = 5.7, y = 1.4$

(d)

1	-8	-6	0	0	900
0	-7	1	1	0	0
0	7	9	0	1	350
1	0	$\frac{30}{7}$	0	$\frac{8}{7}$	1300
0	0	10	1	1	350
0	1	$\frac{9}{7}$	0	$\frac{1}{7}$	50

Optimum $R = 1300$, when $Y = 0$ and $X = 50$.

(e) Optimum $P = \frac{130}{7}$, when $x = \frac{40}{7}$, $y = \frac{10}{7}$

Revision exercise
(page 109)

1 (a)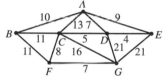

(b) *ADCFG*; order of labelling is *ADEBCFG*; minimum distance is 27 km.

2 Least distance is 163 km.

3 (a) An obvious route is *ADEGFCBA* giving an upper bound of 68 km.

(b) Delete a node, find the minimum connector for the rest, and then add the lengths of the two shortest arcs from the deleted node.

4 (a)

1	-3	-2	-4	-2	0	0	0
0	180	100	**250**	200	1	0	2500
0	5	5.8	6.2	2.5	0	1	100

(b)

1	-3	-2	-4	-2	0	0	0
0	180	100	**250**	200	1	0	2500
0	5	5.8	6.2	2.5	0	1	100
1	-0.12	-0.4	0	1.2	0.016	0	40
0	0.72	0.4	1	0.8	0.004	0	10
0	0.536	3.32	0	-2.46	-0.0248	1	38

(c) The tableau is not optimal as there are negative signs in the top row. Increasing u or v will increase P.

Cost = £2500 and weight = 100 kg

(d) (i) Solution has non-integer values.
(ii) Only whole numbers of tents can be bought.

5 (a) $100v + 200x \le 750$, $5.8v + 2.5x \le 25$, $x \ge 0, v \ge 0$

(b)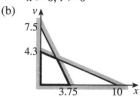

(c) $2v + 2x \ge 9$;

(d) $v = 3$, $x = 2$. The cost is £700 so there will be £50 left over.

6 (a) Using the Nearest Neighbour algorithm from *P* gives *PDABCP* taking 16 minutes.

(b) The minimum spanning tree with *A* removed is 8 minutes. Add the two smallest arcs from *A* giving a lower bound of 15 minutes.

(c) Cooking takes 10 minutes, 3 stops takes $7\frac{1}{2}$ minutes and the delivery, leaving *D* till last could take 14 minutes, giving a total time of $31\frac{1}{2}$ minutes.

(d) In the order *PDABC*, cooking takes 10 minutes, the journeys take 12 minutes, and the deliveries take $7\frac{1}{2}$, so *C* gets the delivery, that is, the start of the delivery, after $29\frac{1}{2}$ minutes.

7 (a)

(b) $(0,0), (5,0), \left(\frac{10}{3},\frac{5}{3}\right), \left(\frac{6}{5},\frac{11}{5}\right), (0,1)$

(c)
1	−1	−2	0	0	0	0
0	1	1	1	0	0	5
0	1	4	0	1	0	10
0	−1	1	0	0	1	1

(d)
1	−1	−2	0	0	0	0
0	**1**	1	1	0	0	5
0	1	4	0	1	0	10
0	−1	1	0	0	1	1
1	0	−1	1	0	0	5
0	1	1	1	0	0	5
0	0	**3**	−1	1	0	5
0	0	2	1	0	1	6
1	0	0	$\frac{2}{3}$	$\frac{1}{3}$	0	$6\frac{2}{3}$
0	1	0	$1\frac{1}{3}$	$-\frac{1}{3}$	0	$3\frac{1}{3}$
0	0	1	$-\frac{1}{3}$	$\frac{1}{3}$	0	$1\frac{2}{3}$
0	0	0	$1\frac{2}{3}$	$-\frac{2}{3}$	1	$2\frac{2}{3}$

(e) After the first iteration $P = 5$, $x = 5$, $y = 0$; after the second iteration, $P = 6\frac{2}{3}$, $x = 3\frac{1}{3}$, $y = 1\frac{2}{3}$.

8 (a) The minimum connector is, *AB*, *AH*, *HX*, *XC*, *XG*, *CD*, *GF*, *FE*.

(b) £25 million

(c) Times to city centre from *A* to *H*, in order, are 15, 20, 6, 9, 12, 8, 4, 7, all in minutes.

(d) Too long from *B*. Cut *AB* and replace it by *BX*. It then takes 8 minutes to get from *B* to *X*, and the total cost is £28 million.

9 (a)

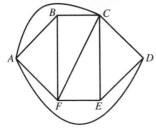

(b) After one pass through Steps 3 and 4, the left list contains *AD* and *AC*, and the right list contains *BF*, *CE* and *CF*. This includes all the arcs, and none is in both lists, so the graph is planar.

(c) There is no cycle which does not cross another arc, so Step 1 cannot be carried out.

(d) There is no cycle, so Step 1 cannot be carried out.

10 (a) Best route, *RSVTXYZ*, takes 156 minutes. The order of labelling is *R*, *U*, *S*, *W*, *V*, *T*, *X*, *Y*, *Z*.

(b) The minimum connector consists of arcs, in order, *RU*, *UW*, *UV*, *VT*, *VS*, *TX*, *XY*, *YZ* with total time 215 minutes.

(c) A solution to the travelling salesperson problem contains a spanning tree plus at least one other edge.

11 (a) *ACBDA*; length 6.7 km

(b) Repeat paths *AC* and *BD*; length is 15.4 km.

12 (a) $10a + 10b + 6c \geqslant 3.5$, $4a + 9b + 7c \geqslant 2$

(b) $48a + 60b + 68c \leqslant 24$, $20a + 2b + 10c \leqslant 5$

(c) $P = 500a + 450b + 350c$ is to be minimised.

(d) The Simplex method has to be used to find a maximum, and $(0,0,0)$ must be in the feasible region.

(e) This condition is $100a + 100b + 100c = 40$. One of the variables can now be eliminated, and the problem solved graphically.

Mock examinations

Mock examination 1 (page 116)

1 Maximise x,
subject to $x + y \leqslant 20\,000$,
         $x + y \geqslant 10\,000$,
      $7x + 4y \leqslant 60\,000$,
         $x, y \geqslant 0$.

2 *AEDCB*, 54

3 (i) $n - 1$

(ii)

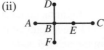

                   53 miles

BD, *BF*, *AB*, *BC*, *CE*

4 (i) There are odd vertices.

(ii) $BC + DE$ is the smallest pairing. *ABCBGFCDEDFEGA*, 2950 km

(iii) Nearest neighbour from *A*: 1500 km Minimum spanning tree, excluding *A*: 1000 km, so $1300 \leqslant$ optimum $\leqslant 1500$

5 (i)
3	3	1	1	1	1
5	5	3	3	3	2
1	1	5	5	5	3
9	9	9	9	8	5
8	8	8	8	9	8
2	2	2	2	2	9

Comparisons   1   2   1   2   5   Total 11
Swaps   0   2   0   1   4   Total 7

(ii) 15; 6, 5, 4, 3, 2, 1

(iii) $1 + 2 + \ldots + (n-1) = \frac{1}{2}n(n-1)$, quadratic

(iv) 200 seconds

6 (i)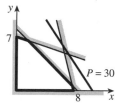

(ii) $x = 7\frac{1}{3}$, $y = \frac{2}{3}$

(iii) $P = 22\frac{1}{2}$ at $\left(7\frac{1}{2}, 0\right)$

(iv) $P = 23\frac{1}{3}$; $r = 0$ so $x + y = 8$,

$s = 11\frac{2}{3}$ so $x + 3y < 21$, $t = 0$ so $4x + y = 30$

Mock examination 2 (page 118)

1 (i) (ii)

P	x	y	r	s	
1	−3	−1	0	0	0
0	2	3	1	0	6
0	1	2	0	1	4
1	0	3.5	1.5	0	9
0	1	1.5	0.5	0	3
0	0	0.5	−0.5	1	1

(iii) $P = 9$. All the elements in the top row are non-negative.

2 (i) Maximise $30x + 90y$,

subject to $x + y \leqslant 48$,

$\frac{1}{2}x + y \leqslant 40$,

$x \geqslant 10$,

$y \geqslant 0$.

(ii)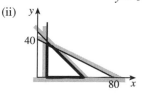

(iii) $(10,0)$, $(48,0)$, $(16,32)$, $(10,35)$

$x = 10$, $y = 35$, £3450

(iv) $(0,0)$ is not in the feasible region.

(v) $x = 16$, $y = 32$, £3360

3 (i) *AFEI* has length 1300 metres

(ii) *AD* 13, *GD* 12, *GI* 10

(iii) There are odd nodes.

(iv) *AGABECDIDEIHGFHEFA*, 7700 metres

4 (i) (a)

1	a	−
2	a	b,d
3	a,c	b,d
2	a,c	b,d
3	a,c	b,d

(b)

1	a	−
2	a	b
3	a	b

(c)

1	a	−
2	a	b
3	a,c,d	b
2	a,c,d	b,c,d
3	a,b,c,d	b,c,d
2	a,b,c,d	a,b,c,d
3	a,b,c,d	a,b,c,d

(ii) Connected if and only if *V* and *W* together contain all nodes.

(iii) Bipartite if and only if *V* and *W* have no nodes in common

(iv) (a) *V* and *W* contain alternate nodes round the cycle.

(b) *V* and *W* both contain every node of the cycle.

Any node of the bipartite graph must contain an even number of nodes.

5 (i) $B \bullet \overset{80}{\underset{C}{\quad}} \overset{60}{\underset{D}{\quad}} \overset{70}{\quad} \bullet A$

210 km, *AD, CD, BC*

(ii) *HCDABH*, 380 km

(iii) *HADBCH*, 370 km

(iv) $50 + 70 + 210 = 330$ km

Glossary

Algorithm	A finite sequence of instructions for solving a problem.
Bipartite graph	A graph with two sets of nodes such that arcs only connect nodes from one set to the other.
Complete graph, K_n	A simple graph such that each of its n nodes is directly connected by an arc to every other node.
Connected graph	A graph such that there is a path between any two nodes of the graph.
Cycle	A closed trail where only the initial and final nodes are the same.
Digraph	A graph with directed arcs.
Euler's relationship	$R + N = A + 2$
Eulerian graph	A connected graph which has a closed trail containing every arc precisely once.
Graph	A set of points (called nodes or vertices) joined by lines (called arcs or edges).
Greedy algorithm	An algorithm where the immediately 'best' step is made without concern about the long-term consequences of this choice.
Hamiltonian cycle	A cycle which passes through every node of the graph.
Minimum connector	A spanning tree whose arcs have minimum possible total weight.
Network	A graph with numbers (called weights) associated with its arcs.
Order (of a node)	The number of arcs meeting at that node.
Order (of an algorithm)	A measure of the 'run-time' of the algorithm as a function of the size of the problem.
Path	A trail such that no node is passed through more than once.

Planar graph	A graph which can be drawn in a plane in such a way that arcs only meet at nodes.
Semi-Eulerian graph	A connected graph with a trail which is not closed that contains every arc precisely once.
Simple graph	A graph without loops or multiple arcs.
Spanning tree	A subgraph which is a tree connecting all the nodes of the graph.
Subgraph of G	A graph whose nodes and arcs are all in the graph G.
Trail	A sequence of arcs such that the end node of one arc is the start node of the next.
Travelling salesperson problem	The classical problem of finding a Hamiltonian cycle of minimum possible weight. (In the practical problem, nodes and arcs may be revisited.)
Tree	A connected graph with no cycles.

Summary of algorithms

Problem	Name of algorithm
Sorting	Bubble Sort
	Shuttle Sort
Packing	First-Fit
Minimum connector	Prim's
	Kruskal's
Shortest path	Dijkstra's
Route inspection	Chinese Postman
Travelling salesperson	Nearest Neighbour
	Lower Bound
	Tour Improvement
Linear programming	Graphical
	Simplex

Index

The page numbers given refer to the first mention of each term, or the shaded box if there is one.